ISLANDS OF BLUE WATER

KEITH ROBINSON

Islands of Blue Water

ADLARD COLES LIMITED

© Keith Robinson 1968
First published in 1968 by Adlard Coles Ltd.
3 Upper James Street, Golden Square, London W1
Printed in Great Britain
by Ebenezer Baylis and Son Ltd.
The Trinity Press, Worcester, and London
SBN 229 97363 9

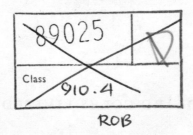

To the memory of Robert Somerset

CONTENTS

LIST OF PLATES

LIST OF DIAGRAMS & CHARTS

Maps drawn by Jean Stewart

PREFACE

This book is the story of the beginning of what I hope will be many years of living in a dream come true. It is the story of the sailing of my boat, which is also my home, in the Caribbean, the Atlantic and the Mediterranean, and of places and people brought into my life by this boat.

The book is not written as a technical guide to sailing, pilotage or travel; I have not the skill or knowledge to attempt such a thing. It is written to tell of a way of life which may interest some people and which some might even wish to follow.

It is a way of life which has brought much pleasure to me.

Puerto de Andraitx Spain 1964   Keith Robinson
and Volos Greece 1965

How it all began

I was shivering as I stood at a Welsh dockyard in a cold February night. I looked up at the white ship that was to take me to the West Indies and to the small yacht waiting for me there in the warm sun.

Then I took my first step onto the gangplank and into a new life. It was not only the cold which was making me shiver, but excitement and doubt over what I was starting to do. I hoped that it was the right thing, because right or wrong, I was now committed to it.

As a boy of ten I had been caught by a love for sailing. My parents and relations had no interest in boats, nor had any of their friends except for one man. We were staying with his family one summer and he asked me if I would like to use his sailing dinghy. 'I don't know how to sail.'

He got up and said, 'Well, I haven't got much time now, but anyhow, come along. I can show you the basis of sailing in twenty minutes. Then if you like it you can go on learning for the rest of your life.'

I did like it, and for the rest of our visit I sailed that dinghy for every minute that I could find. Willie Woellwarth had given me the best gift that I can think of, a start in sailing, and his help whenever it was possible. Unfortunately, it was not possible very often. I used to get what sailing I could, often by hiring dinghies or half-deckers, and I read everything about sailing that I could buy or borrow.

By the time I was twelve I was sure that the best possible life was one of sailing a small boat in warm waters. Later I got nearer to this by inspiring

four other boys with some of my eagerness, and getting them to share
the cost of chartering a 30-foot sloop in the lakes and rivers of the Norfolk
Broads. It was not sailing in warm waters, but at least it was sailing. We
used to go in early April. We sailed in cold rain, in frost, and even in
snow. This did not kill my wish to sail, and maybe it made the idea of
warm waters still more attractive.

For some reason my keenness to sail did not make me want to take to
the sea as a profession, and when I left school I went into the army, a year
before the 1939–1945 war began. A soldier can often get a lot of sailing,
but I was not in the right places and I got very little. I started my soldiering in the driest part of India, and then spent a year in the most barren
mountains. Then the war came, and my life took me to various deserts for
three years, to England for six months when there were other things to do
than sail, and finally for a year to Burma.

In spite of all this, I managed to get some sailing in places as far apart
as Egypt and Burma. After the war spells of soldiering in England,
Germany and America brought some good weeks of sailing. I liked the
American sailing best. The waters of the Maine coast were too cold for
me, but the warm waters of Chesapeake Bay and of the islands south of
Cape Cod convinced me that I wanted to live in a boat.

The sailing that I wanted was more than just having odd weeks in other
people's boats or in chartered yachts. I wanted to live in a boat of my own,
free to move her where I wished. This meant having lots of time and some
money. I could make the time, but money was a problem. I would have
to find enough capital to buy a boat and enough income to keep both the
boat and myself. I could imagine that once I had a boat it might be possible to earn money by chartering or carrying cargo or passengers, but I
was not keen on the idea. To make money I should probably have to own
a large boat needing a crew. I wanted to be free of having a crew, amateur
or professional, except when I chose, and I wanted to be able to enjoy my
sailing without being forced to work for it. I was too young to be able
to retire with enough pension to keep myself, let alone a sailing boat as
well. No legacies were going to come my way, and there seemed no good

solution to the problem except a long period of waiting. That was not what I wanted.

Then just before I was forty a solution appeared, or rather two un-expected events which together produced a solution.

I was sent to Malaya to command an armoured car regiment, which brought promotion. Soon after that a retirement scheme was produced to give those who retired earlier than normal the full pension of their rank, and a good capital grant, in my case £5,500. It would have taken me many more years in the army to earn that much pension in the normal way and to save even a small part of the capital. The chance was too good to miss, and I applied to retire.

Applications and refusals and more applications went on for three years, but finally I retired. I went back to England by the Pacific, taking five months over the journey. My duty had brought me half way around the world so it seemed a pity to return the same way, and I wanted to have a look at areas which seemed possible bases for my sailing life, which was now coming within reach.

The islands around Malaya, Borneo, and the Philippines were beautiful, with perfect climate for most of the year, but for political and piratical reasons they were not a sound proposition. As an alternative, I knew something of the Mediterranean, and it seemed a likely choice. There were two other areas of sea which were attractive and of which I knew nothing. They were the Caribbean and the South Pacific. I wanted to see them and to find out their problems.

My plan was to fly to Australia and then to Fiji where I would take a flying boat to Samoa. Then again by flying boat to Tahiti to spend August there, hoping to visit the Tuamotus and some of the smaller islands of the Pacific, perhaps by trading schooner. From there I was to go by ship through the Panama Canal to Martinique, with two months in the West Indies and then a month in Virginia before reaching England by Christmas time.

After that I wanted two months in England to arrange my retirement and make my farewells, and then spend the rest of the spring and summer wandering about Europe by car. As my last venture on land I hoped to find a boat to buy before the winter. From then on I had no plans, but I thought I might spend that winter and the next summer or two in the Mediterranean, and then perhaps sail to the Caribbean.

My plans worked well until Tahiti. There I found it very hard to arrange movements to the distant islands, and it was impossible to fix a return date. I had not allowed enough time. The best I could do was to visit Les Iles Sous Le Vent. I flew to Bora-Bora.

There by chance I met some Americans from a large motor yacht. They invited me to go to Seattle with them, starting in three days. I began to say that my journey was all arranged and that much as I should like to go, it was impossible. Then I saw that for once in my life I had time to spare, and the only person bound by the arrangements was myself. So I left with them.

We moved fast northwards, stopping at some of the small atoll islands, heading for Honolulu on the way to Seattle. Our passage was too fast and our stops too short to get more than fleeting impressions of the islands and their peoples, and our contacts were only superficial. Two hundred miles from Honolulu the boat developed a bad leak, and we made harbour after three days of strain and hard work. The boat had to be docked for repairs, putting an end to the Seattle part of my journey. This was in early September. I decided to go to Virginia first, and then to spend November and part of December in the West Indies.

This change of plan was a lucky one for me, because without it I would not have met the owner of what is now my boat. Nor would I have met him, I think, if it had not been for a small red dog in Malaya, named Jessica. She had been picked up, a mongrel puppy, in the middle of a busy street, and brought to me to be kept for one night. She stayed with me for three years and became a dog of character. Jessica formed a close friendship with the enormous Alsatian belonging to my General, and this attachment did much to make me a frequent visitor to Flagstaff House.

Before I left Malaya for my eastward journey, the General gave me an introduction to a friend of his in the West Indies, living in St Lucia, two hundred miles north of Trinidad. I was invited to stay with him on my way through the Caribbean. Jack Harrison's invitation and my change of dates at Bora-Bora brought me to St Lucia just as Alan Keith arrived by air from England to get his boat ready for the Caribbean season of winter sailing. But for this chance meeting I would not have found *Leiona*.

I was not then looking for a boat to buy. At that time I was still trying to learn what was the best sort of boat for my new kind of life. Whenever I could, I was visiting boats and talking to their owners, and probably being a general nuisance to them. Before I had left Malaya my idea of the right boat, small enough to sail alone and yet big enough to have two or three others with me when I wanted, had been a ketch about 35 feet long. The two masts would split up the sail area and so make any one sail smaller and easier for me to handle on my own.

During my Pacific wanderings this idea had been changed by the views of several people whose varying opinions I valued. These views were quite independent, but they all pointed to the same general idea; a boat about 40 feet long, preferably with the cockpit somewhere amidships. This feature, a strange one to me, would give safety and comfort at sea, and some degree of privacy for living aboard, by separating the sleeping cabins. Everyone said that a ketch might be the easiest to handle, with its small mainsail, but that as long as there was roller reefing a cutter was no great problem, even for a single-hander.

At St Lucia several sailing boats were anchored in Vigie Cove, near Castries, most of them fairly large. One day I was invited on board a white cutter anchored at the mouth of the cove. I had wanted to see her because she was smaller than the others, and more of a size that interested me. She was Alan Keith's boat. He had sailed her from England two years before, and came out each winter to live in her. He had a business in London in partnership with his brother, and they had the excellent arrangement of each taking a half year's holiday.

I liked his boat. She was strong and good-looking, well kept and comfortable, and gave a feeling of confidence and ability. On a waterline of 35 feet she was 41 feet long over all, with $11\frac{1}{4}$ feet of beam and only 5 feet of draft. She was a solid boat, with lots of room inside her. The accommodation seemed right to me. There was a small stern cabin of two berths. Forward of that was a small and deep cockpit with the engine below it, the auxiliary generator in a locker on one side, and warps and fenders on the other. Twin fuel tanks were below these lockers, which formed the cockpit seats. Next came the saloon, a big one with two settees which could be used as bunks, lockers, a chart table and shelves, and a big swinging table in the middle. The freshwater tank was below the table. Forward of the saloon the galley and gas stove were to port and the lavatory to starboard. Then came a large two-berth cabin and a very big forepeak in which were sails and gear and a gas refrigerator.

Her gear was good and strong. The mast was stepped well aft, almost amidships, and with the sail plan split into a fairly small roller reefing mainsail and two large headsails she looked easy to handle. There were winches for sheets and halyards and good big wooden cleats. In every way she seemed just the sort of boat I wanted for myself, and I was very sorry to leave her.

As I got into the dinghy to row ashore I said, 'Good-bye, Alan, and thank you for letting me take up so much of your time. I think you have a fine boat here, and I really do envy you. Anyhow, good luck, and I hope you have many happy years of sailing in her.'

His answer surprised me. 'Yes, she is a good boat, and I am very pleased with her. But with my growing family we find that we need a bigger boat, and I have ordered a larger one to be built in Holland.'

'You seem to be starting a fleet. How are you going to cope with two boats at once?'

'Well, I shall have to sell this one. There isn't much of a small boat market in the Caribbean, so in the spring I am going to have her sent back to England on a steamer. She will be easy to sell there.'

I found that my mind was in a turmoil of thoughts and ideas. While I was trying to sort them out I heard my own voice saying:

'Alan, I would like to buy this boat. What is your price?'

I was astounded at what I had said. So was Alan. He had not thought

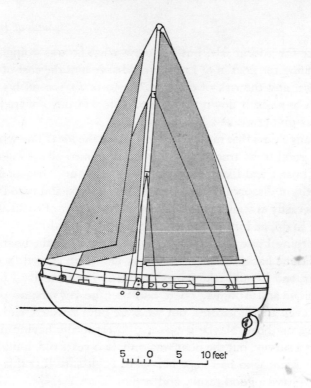

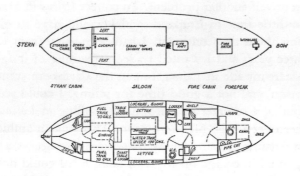

Sail plan and accommodation and deck plans of *Leiona*

about a price for a local sale, but he knew what he was going to ask in England. Selling the boat in St Lucia would save him the cost of shipping and insurance, and the risk of damage. He took £1,500 off his intended price, which brought it down to £5,000. This was just covered by what the army was giving me as a leaving present.

I spent many hours that night thinking about the idea. The whole thing seemed too good to be true, and too easy. It seemed all wrong to decide on the first boat I had liked, or even thought about. I had been told to expect months of disappointing searches and surveys, and here I was buying a boat as easily as walking into a shop to buy a pair of socks. It seemed a silly thing to do, and maybe a rash one. I decided to do it.

In the morning I saw Alan and said that I would buy the boat. It suited us both if I came back in March to collect her. Alan agreed to make out a list of gear, and I could take what I wanted. The next day I flew on to visit Martinique and Antigua. There had been no lawyers, no papers, no survey, not even any signatures, but we were both satisfied, and in March I would have my boat. Maybe it was a risk, especially having decided to buy without a survey, but the boat was only two years old, built by David Hillyard for Alan, who had kept her well. I felt sure that this hasty act was going to have a good result, and it did.

I had given myself another problem. To collect *Leiona* in March would leave me very little time in England, and I would have to give up my plan of driving about the Continent. I had been away from Europe for more than three years and I wanted to see something of it again. I was impatient to start my life in *Leiona*, but in the Caribbean summer is the 'hurricane season' and not a good time for sailing. I could see only one way of combining a summer in Europe with sailing in *Leiona*, and that was to sail her over to the Mediterranean. This was an ambitious idea, giving me an Atlantic crossing as my first attempt at deep-sea sailing, or at any sailing at all in charge of my own boat, but I could not see what else to do. The idea was certainly a possible one, even if frightening, and I decided to try it.

There was a lot to do, and very little time to do it. After a short visit to

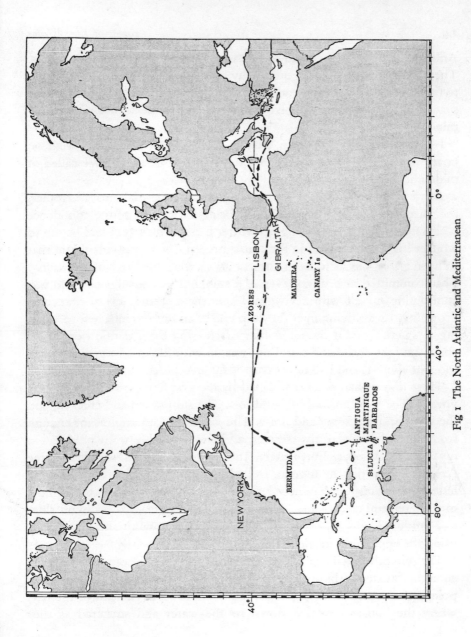

Fig 1 The North Atlantic and Mediterranean

Antigua I flew to England and had six weeks there. They were busy weeks. There were friends to see, gear to buy, and finances to settle. I managed to fit in a few days at Commander Lund's school at Torquay to learn about sextant navigation, and I hoped that all he taught me would stick in my mind.

I had booked a passage in a 'banana boat', and in February I went on board her in a Welsh harbour, feeling cold and doubtful. We sailed at midnight.

As we went south from Wales, through the Azores and into the tropics, the warm sun and blue seas began to bring back some of the confidence which the grey cold of an English winter had taken away. I had begun to wonder if I had let myself in for a mad project. Dr Johnson had said that to be in a ship was as to be in a prison, with the added risk of drowning. I had committed myself, not even to a ship, but to a small yacht, and was confronting myself with a voyage where there seemed every chance of drowning. I was diving in at the deep end of an untried idea before learning to swim. Even if the difficulties of crossing the Atlantic were overcome, I might find that I could not put up with life in a boat. I had only thought about it and had never tried it out except for very short spells.

These doubts had not been helped by the first few days of the steamer voyage. The weather was cold and bad. The ship rolled and creaked, and the wind brought spray and noise. The waves looked unpleasant enough from the high decks of the steamer, and I wondered how I would feel in my own small boat. Fortunately, things got better as we cleared the Azores, and so did my morale. I filled in the time writing long-overdue letters and trying out my new sextant and navigation tables to see how much I remembered of my Torquay instruction, when the Devon skies had produced only rain with no horizon or sun to allow me to practise using the apparently magic instrument.

The ship called at Trinidad and Grenada, and then we came to St Lucia, entering Castries harbour early in the morning. We rounded the southern point and steamed slowly towards the white town at the end of the bay, where the houses crowded down to the water and scattered as they stretched up the green hills all round us. *Leiona* was the first yacht we passed, anchored at the mouth of Vigie Cove.

The steamer docked beside the long iron sheds filled with piles of banana stems, some already in their long plastic bags, some in mounds of green and yellow, and still others being unloaded from the trucks to the scales. It was a great contrast to the grey silence of the Welsh dockside. Here was noise, colour and bustle. Tail-boards rattled, brakes squealed, and engines roared as lines of trucks, their sides gaily painted with water-falls and mountains, backed up to the sheds. The heavy stems of bananas were swung off in an even flow onto the shoulders of the noisy chain of carriers, men and women in ragged working clothes of faded colours, singing and laughing as they carried and flung their loads.

Alan met me at the dock, and took me across the harbour in his dinghy, after finding a truck to bring my gear around to Vigie. He had to get back to his business in England, but stayed five days, and was kind and pains-taking in his guidance to me. In exchange for the formal bill of sale I wrote the biggest cheque of my life, and the boat was mine. There was a lot for me to learn and to do to keep *Leiona* as carefully as Alan had, and I was sorry when he left.

I planned to start the Atlantic crossing at the beginning of May, and to spend the whole of April getting ready in Antigua, in English Harbour. This left me March in which to get to know about my boat. My idea was to sail in the Windward and Leeward Islands as far as I could and as hard as I could in order to try out the boat and myself. I wanted to practise sextant work, taking sights where I could check my fixes with the land, and I wanted to test the boat and myself for weaknesses. If anything was going to break, it was better for the break to happen before rather than after my start to cross several thousand miles of ocean. And I wanted to see how I could handle the boat and look after myself when sailing alone.

I was not really worried about *Leiona*'s fitness. I felt that a well-kept boat, two years old, was probably in better condition than a new one, as any faults would have been found and put right. The engine was a Parsons Pike, a four-cylinder Ford diesel engine, and it ran well. The rigging was all good except for the main halyard and the topping lift, which were easy to change. The mainsail was a new one of Terylene, and the cotton jib and staysail looked all right for one more season. There was a light cotton Genoa, and a pair of large staysails to fit to twin booms for down-wind sailing, when she would steer herself by sheets leading to a temporary tiller. Normally the steering was by a wheel.

I had some doubts about three of the fittings. One was the mast tabernacle, a narrow, upright, iron box on deck, bolted through to the beams, in which the mast was stepped. It did not look massive enough, and there was no supporting rod or metal strengthening below the deck. However, Alan and others assured me that there was no need to worry; David Hillyard had been using this system for about fifty years, and it was well tested. They were quite right.

The fitting at the top of the rudder post, the steering quadrant, did not seem to have been fixed strongly enough. It was a 90° 'wedge' of iron plate from which the steering cables led to the wheel in the cockpit. It was clamped to the rudder post and held tight by a rivet in each arm of the clamp. There were two bolt holes on each arm, but these were for bolting on an iron tiller when the twin staysails were used for the down-wind self-steering rig.

The other fitting which looked doubtful was the masthead attachment for the standing backstay. It was a narrow bar welded to the masthead band, projecting horizontally for about six inches. There were no running backstays, so all support the mast had against forward strain passed through this fitting. Again, I asked experts about these things, and they all told me not to worry. The worries came later.

A more immediate problem was that of a crew. I could not afford to pay and keep even one paid hand, and, anyhow, I did not like the idea of having one in so small a boat. I felt sure that I could handle *Leiona* by myself in my island wanderings, and even across the Atlantic, if necessary, but I would rather have somebody with me for the long journey. The biggest difficulty in collecting a crew seemed my own lack of experience. I did not feel that I could fairly ask anyone to come with me when I knew so little myself. Nor did I want to be under the care of an expert, and just be a passenger in my own boat.

There did not seem to be a good answer to the problem, and I was resigned to going alone. A course from Antigua to Gibraltar passing near Bermuda and through the Azores should give a down-wind passage most of the way, when the self-steering gear could do the work and let me rest. Even with winds on the beam or the bow I thought that *Leiona* would sail herself well enough to free me from steering for most of the time. In any case, if alone I would be able to carry so much food and, more important, water, that there would be no hurry to reach land, and by not having to drive myself or my ship, I could ensure that there would be enough time for sleep.

Two people had already asked to come as crew, but the arrangement had been very nebulous and I had heard nothing from them since our first and last meeting in December. It was now March and there was still no letter from them. Our discussion had been very brief. While in Antigua in December, before returning to England, I had stayed for a few days at Nelson's Dockyard in English Harbour. One evening the owners of *Sea Otter*, an American ketch from the Virgin Islands, asked me on board. They were giving passage to a young Frenchman and his wife, both journalists on their way to Yucatan. Michel Peissel was a great traveller and had just come back from an expedition in Nepal and Tibet. I had known the family of Charles Bell, who had spent a long time in Tibet, and he had written and published a Tibetan Grammar. Michel had found a copy of this Grammar in a second-hand bookstall in Paris and it had been the cause of his journey to Tibet. We carried on an enthusiastic conversation about Tibet, the Bells, India and Nepal, without once mentioning boats, an almost impossible thing to do in English Harbour.

Early next morning *Sea Otter* sailed for Martinique. No one else in the Dockyard was stirring and I handled their lines. Michel must have asked about my sailing plan after I had left the night before, and had been pondering over it. He was standing in the bow. As the boat drew away stern first, he called out to me.

'Are you really going to sail to Europe next summer?'

'I hope so.'

'Can I come with you?'

They drifted away. No time for discussion.

'Yes!'

Further away still.

'Can Marie-Claire come, too?'

Almost out of hearing.

'Yes!'

'Where shall we meet?'

Nearly gone.

'Sailing the first week of May. You come here the last week of April!'

'All right. Good-bye!'

Then they were away, without any address or any means of my finding them. In spite of the casual form of our agreement, I felt that they would come, and as it was at their own wish and not by my suggestion, I did not feel guilty about them. I had hoped to find a letter at St Lucia. Maybe there would be one at English Harbour. I wondered how they were faring among the Indians of Yucatan. I could see myself being deprived of a crew by such disconnected happenings as scalping, estrangement or pregnancy. Like Mr Micawber, I could only hope that something would turn up. Anyhow, I took on some water and food and set off for a month of sailing on my own.

CHAPTER TWO

Caribbean start

March in the Antilles was a wonderful month.

I wandered among islands of flowers and green hills, brilliant trees, white beaches and sheltered anchorages where the waters varied in colour from deep blue to aquamarine, turquoise, palest green, and almost white. It was a month of good sailing. Sometimes there were hours of calm under the lee side of the hills, but more often the wind was strong enough for a few rolls in the mainsail. Sometimes a rain squall would send the boat racing along with eased sheets. In the open passages between the islands the seas were big, pushed across the whole Atlantic by the Trades, and the winds were strong and steady.

I found *Leiona* easy to sail single-handed. With a little trimming of the sheets she would steer herself on almost all points of sailing. I lost some of my misgivings and began to look forward to the Atlantic crossing, even if alone.

From Vigie Cove I started southward, passing Marigot Bay and the tall peaks of Les Pitons. My first sail in my own boat was a thrilling one across 30 miles of open sea with three rolls in the mainsail. Near St Vincent the sky was turned crimson, orange and purple by a flaming sunset. It then faded into an ink-blue arch with bright stars showing between patches of racing rain clouds.

I chose a place in the south part of Kingstown Harbour, and anchored in deep water between the riding light of a big trading schooner and the windows of some small houses on a steep cliff. For the first hours of the

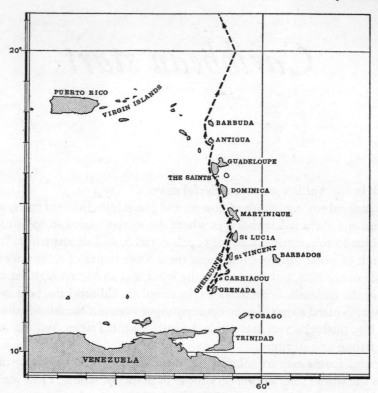

Fig 2 The Caribbean

night there were angry gusts of wind which made the boat snub and swing on a tight chain. I hoped that the anchor would hold, as the cliffs were very close.

Before morning the wind died away and I woke to find a silent dawn breaking. I also found that the boat was still close to the cliffs, but was now facing towards them instead of away, which seemed impossible. When I looked over the bow into the clear water, just becoming light, I saw what had happened. It was my first experience of anchoring in tideless waters. When the wind dropped, there was no current to move the boat, and she was left, held by the weight of the chain, pivoting on the short length below the bow, and not on the anchor, many fathoms away in the deep water.

I sat for a while in the cockpit watching the dawn light fill the sky and fall on the slopes of the hills around the harbour. There were no sounds and no movements except the shadows. Then a door opened from a small white house on the edge of the cliff above the boat. A young woman appeared. She carried a bowl and a towel and wore a red cloth around her body. Her skin was a pale golden-brown. She picked her way down the cliff-face, reaching out with her long legs and balancing with her arms. Near the foot of the cliff there was a cleft in the rock and from it trickled a tiny waterfall. The girl stopped beside it. She put her bowl and towel on a ledge, then unfastened her red cloth and hung it on a stone. She stood naked into the waterfall, letting the stream run first on to her throat and shoulders, then guiding the flow with her hands over her breasts and body and down her arms and legs. She then took her soap from the bowl and washed herself, leaving her head dry, the lather white against the gold of her skin. She filled the bowl and poured water over herself, washing away the soap until she stood gleaming before the rising sun, like a Tahitian vahine, except that her hair was short and curling. She dried her body, wound on the red cloth and climbed up the narrow track. The door closed and the cliff was empty. I was left with the silence and a memory of beauty to take with me when I sailed.

With the sun came a light wind, and I let it take me out of the harbour. From St Vincent southward to Grenada stretch the Grenadines, a chain of small islands like a necklace laid out on a cloth of blue sea. Sailing among them was beautiful and peaceful. Sometimes a few houses were in sight, sometimes another yacht, but more often there were only barren islands with bright colours of rocks and sands and waters, and the white wings of terns and bosun birds. Even the names were beautiful, like sounds of music . . . Bequia, Mustique, All-Awash, Cannouan, World's End, Frigate Island, Mayero.

Tobago Cays (called Keys) was confusing as it was in the Grenadines and nowhere near Tobago. It was a small cluster of barren and deserted islands, joined and separated by fascinating channels, lagoons, and reefs. I found the main channel filled by four big yachts, so I went on through it to pass over a shallow bar and turn south down a narrow inlet to a small lagoon, lonely and well-sheltered. There was a slight swell, and I sat in the sleepy boat, watching and listening to the heavy surf breaking on the reef with the open Atlantic beyond.

Carriacou is the biggest of the Grenadine Islands and Hillsborough, its port and capital, is a town and not a village. I anchored there and went ashore to get clearance for entering the Grenadines and Grenada. It was surprising and sometimes irritating to find that in the British West Indies, even with federation about to take place, as much formality was needed in passing from one island to another as in passing between two foreign nations. It was necessary to be cleared for entry by three separate authorities; Customs, Immigration and Health. No difficulties were ever raised, but a lot of time could be wasted.

The Health officer of Carriacou, Dr Charles, was very kind to me. He drove me around in his car and showed me the island. It was flatter and drier than St Lucia, more like Sicily than a Caribbean island. The crops were cotton, maize, pigeon-peas and mahogany. There was also a thriving commerce in the smuggling of wines and spirits.

The view from the highest point, the site of the doctor's house, was magnificent. From there we looked down upon all the islands of the Grenadines, and on to the coral reefs and lagoons of the eastern side. The height gave contrast to the colours, and the waters became a mosaic of blue, green and opal. Far away to the north-east was a solitary island formed by a single white rock. It had the look of a square-rigged ship and was called Sail Rock. There are stories that in Nelson's days it had often been mistaken for an enemy frigate, and that one captain fought a hard battle with it during the night, mistaking the echoes of his own guns for the broadsides of the enemy.

Dr Charles drove me down to a little village on the eastern side of the island, with the simple and appropriate name of Windward. In its lagoon

1. Reduit Beach, St Lucia
   Photo: Gerald Ross

2. *Leiona*
   Photo: Robert Riddle

3. Nelson's Dockyard, Antigua. *Photo: BOAC*

4. San Miguel, island near Ponta Delgada, Azores. *Photo: Portuguese State Office*

harbour was the big red schooner *Amberjack* whose owner took us to his house where we drank good French brandy which had probably been brought from Martinique. Then we walked around and looked at the boats. Hauled up on the beach, rigged for jib and spritsail, were some small open sloops. In a grove of the poisonous manchineel trees was the half-built hull of a schooner on which work had stopped when the money ran out.

The method of building these schooners was interesting. No use was made of a shipyard; all the work was done on the nearest bit of suitable beach to the owner's house. The boat was built upright in the normal way, but for launching it would be let over on to its side and hauled down to the sea on rollers under the bilge. The keel was usually a good piece of timber from British Guiana, properly sawn and squared, but frames, ribs and planking were locally cut and very roughly finished, often with the bark left on. The hull had a long, flat run to the counter which ended in a raked transom. These schooners are fast boats and sail well, but there are many tales of losses in bad weather, when they have been known to open up and sink.

We drove back to Hillsborough in the evening and I went out to *Leiona*. There were seven schooners and two yachts in the harbour so I sailed around to Tyrrhel Bay for the night and had it to myself. A small boy rowed out in a very heavy boat with a basket of limes, which were welcome. He also offered to pick me some oysters. There may be other places in the world where oysters grow on trees, but Tyrrhel Bay is the only one I know. There the oysters fasten themselves in thick clusters to the mangrove branches and can be picked like bunches of grapes. They are very small and not really worth all the trouble of opening, except for the novelty. Fish were easy to catch there. By holding a light close to the water, one might scoop up with a net the inquisitive mass lured to it. I tried this method with success and gave my catch to the lime seller before sailing in the morning.

From Carriacou I sailed south to Grenada, past the big rock with the odd name of 'Kick 'em Jenny', where the currents and waves built up

2

some ugly seas. It was good to come under the shelter of the bigger island, where there was enough wind to give a pleasant sail close in to the shore.

It was afternoon when I sailed along the western side of Grenada. The lee sides of the islands were never at their best in the mornings, when the eastern sun threw the long shadows of the mountains over the sloping forests and masked their colours. Now the clear light from the west brought into sharp relief the ridges and valleys, and among the greens of the hillsides the flamboyant trees made bold splashes of scarlet. A few white houses with roofs of red or grey added colour, and scattered cliffs and beaches lent edgings of brown, ochre and yellow to the white lace of the breakers. The sound of the waves and the sweet scent of the land reached out to the boat.

St George's, the capital of Grenada, was like a Corsican port. Two tall, steep peninsulas reached out to embrace the harbour. One of the headlands was topped by an old fort, now a hospital, and the other by the white block of the Santa Maria Hotel. Inside, the whole was divided by two small, straight promontories into three separate bays, one being the waterfront of the business section of St George's, another where the steamers docked, and the third a large lagoon behind a coral reef. It was a pretty town with steep, narrow, winding roads and houses piling on top of each other, with walls of white, pink, red and yellow. Bougainvillaea, hibiscus and flamboyants added colour to the green of the trees and lawns, and the background was a steep slope up to a high, straight ridge crowned by another old fort.

A narrow passage into the lagoon had recently been made through the reef, and provided a really good anchorage near the little yacht club.

Grenada was my favourite island, and the one I got to know best, thanks to John Slominski. John was a Pole who had fought in the war in the Free French forces and in the British Army. He was a doctor, and when I came to Grenada he was the government medical officer for the island. He was very good to me. He took me to stay in his house on the mountain-side, looking westward over the sea. He drove me all over the island, from the cliffs and mountains of the north and centre to the dry, flat land in the

Caribbean start

south. He took me not only to the estates of planters but also to the houses of the poor people of the island and told me of their owners and their customs. He had a wide knowledge of people and countries in peace and war, and the wise judgement and philosophy of a doctor. From our first meeting I had felt an admiration and liking for him because of his amazing likeness in looks, speech and thought to my own father, who had also been a doctor.

John took me sailing in his little sloop, *Janka*, and that was a good experience. We went to Carriacou and back, and for the first time I saw the fullest use made of steady winds. We set off in the afternoon, and John said:

'We shall have to do without a compass, because I have lent it to the Navigation Class, but we shall be all right.'

As we were about to sail to an unlit island, mostly by night, I hoped that he knew what he was doing.

It was a windy day with some rain-squalls. After a while, John changed the Genoa for a working jib and we settled down for a long sail to windward. Our course was taking us a little west of north, increasing our distance from land. Soon after sunset we tacked and headed for 'Kick 'em Jenny', which we could just see eight miles away. I thought that, in a cloudy night with no moon and few stars, we would have a difficult time finding one small rock in the dark. Some booby birds were flying in the right direction, but we would not see them much longer. John's instructions were very confident:

'Keep as close on the wind as you can. In about two hours you will hear the waves breaking on "Kick 'em Jenny". Sail straight on and you will see the surf. When we are close to the rock we will tack and head for Carriacou.'

And that was exactly what happened. The next lot of instructions were on the same lines:

'Keep as close on the wind as you can. In about two hours you will see a pair of rocks sticking straight up in the sea. They are called The Sisters. We will go right up to them, then tack and sail into Tyrrhel Bay.'

It happened just like that again.

I was steering when we came to The Sisters, and woke John, because I was not sure if they were two large rocks still far away, or small ones much closer. We tacked within a few yards of them and sailed between two low headlands into the bay. It was too dark to make out the shore line, but John sailed on until the sound of the shallow surf told him that the beach was close. We then anchored in two fathoms and went to sleep. The next day we had a gentle sail back to St George's in light wind and warm sun.

I was sorry to leave Grenada, because I was leaving friends, but I wanted to have *Leiona* hauled out on a slipway and new antifouling painted on, before the beginning of April, when I planned to be in Antigua.

Grant's yard in Martinique had a place for me on their slipway, and I wanted to be there as soon as possible. There were some known places to see again, and some new ones to visit. The wind and the weather held good for the northward passage. Grenada and 'Kick 'em Jenny' dropped astern, then Carriacou, Union, Cannouan, and the rest of the Grenadines to Becquia. Then came Kingston and Cumberland Bay in St Vincent and, at last, Les Pitons of St Lucia stood up bold ahead. I called at Castries to make my farewells to Vigie Cove, and especially to Bert and Gracie Ganter and their helpful shipyard. My next course was from Pigeon Island to Diamond Rock—H.M.S. Diamond Rock to the Royal Navy, whose ships dip their ensigns to it in honour of its fight against the French. I sailed close to its high steep sides, and thought of the young officer who had managed to get heavy guns up the cliffs. It seemed an impossible feat, even for those days of wooden ships and iron men.

One of the biggest joys of sailing in tropic islands was the life that was always appearing in and above the water. It was hard to sail for an hour without seeing something of interest in the sea or in the air. Porpoises were the most rewarding of all, because of their size and almost human expressions, but there were many others to catch the eye. Coveys of flying fish were so common that they ceased to be a surprise. Sometimes the blue of the sea was cut with silver, either the single flash of a leaping tuna or swordfish, or by the spatter of a school of jackfish or dorade. The air also had its life. The largest birds were the heavy pelicans, clumsy when on a rock or when landing, but graceful in their effortless, gliding flight, or the big frigate birds with their angled elbows. The smaller ones were the terns and the needle-tailed tropic birds which, in flight, often took on a green colour, reflected from the bright turquoise of the clear water above the coral sand. The smallest of all, not found at sea, but to me the most impressive of the Caribbean birds, was the humming bird with its blur of hovering wings and its feathers of shining blue and green.

My short time in Martinique was very happy. Felix Boisson, a splendid and volatile colonel retired from the French Colonial Army, took me in charge and occupied every hour that I could spare from Grant's yard. He was a man of unfailing energy, in spite of having lost one leg and half an elbow in two wars. His losses did not keep him from swimming, hopping down to the waves, while leaning on my shoulder, nor from driving his car with great speed on the mountain roads, gesticulating to aid his descriptions, and complaining about the condition of the brakes.

I wish that I could have had more time in Martinique. There was a great contrast of city, town and village; mountain, coast and plain. All the coast-line was beautiful, the steep, rocky corniches of the west and north, the pretty little bays and islands of the east with the coloured corals and lagoons, and the beaches of the south with white and black sands, glisten-ing with mica. The mountain mass of the north was impressive, but for me the most pleasing beauty was in the lower hills and rolling plains of the south and centre. The huge sweeps of sugar-cane fields were then without their high plumes, like white feathers, but even so, were beautiful. The wide valleys were pretty with their varied fields, woods and copses of light and dark green, while the waving branches of the hedges, grown from a willow-like tree called glyciandrina, outlined the fields. Wherever this willow was used for making a fence, even when just driven in as a stake, it began to grow until each fence-line became a flourishing green strip of woodland.

I had sailed into Fort-de-France, the capital of the island, after dark, creeping in among the yachts and small boats anchored in front of the

Savannah, below the fort. It had not been easy, because only a few of them had riding-lights, and even those had been hard to see against the bright lights of the city. In the morning I woke to see the sun rise over the walls and gun-embrasures of Fort St Louis. We were lying in an open anchorage, rolling heavily in a slow swell, near the big black schooner *Freelance*. The city presented a picture of white and yellow buildings, palm trees and bougainvillaea, the roads busy with cars and people.

A wide promenade ran across the front, and behind it was the wide grassy Savannah, an open square with eight tall palm trees surrounding the statue of Josephine, Empress of France, born in Martinique. She gazed south, across the beautiful bay towards Trois Ilets and the place of her birth. My visit to her home, La Pagerie, left the most pleasant of all my memories of Martinique.

The drive around the bay to La Pagerie was beautiful, with the sea and swampland on our right, and small hills and fields on the left. We visited the church in the pretty village of Trois Ilets and then turned inland up a narrow sandy lane to the sugar estate where Josephine Tascher de la Pagerie had lived as a child, before her marriage to de Beauharnais. It was a charming little farm lying in the shaded valley of a clear stream. A tall broken chimney stood near the ruins of a big pressing mill which had been turned by oxen, and could be reached by a footbridge from the small house. The house, white and covered with bougainvillaea, was the holiday refuge of Robert Rose-Rosette and his charming wife, Simone. Part of the house had been made into a small and intimate museum, housing letters, pictures and souvenirs of Josephine and her family and of Napoleon. I had the feeling that it was still her home.

Felix Boisson filled my evenings with introductions to the people of Martinique, white, black and brown, and to its food and drink. The people were delightful. I met a wide selection, ranging from the Commander-in-Chief and the Rotary Club, to cane-field workers and fishermen, and an unforgettable tall man, mending his nets on the beach at Ste. Anne, coal-black and scarred with wounds from Verdun.

The colourful people of Martinique seemed happier, better dressed and better educated than those in the other islands I had seen. They spoke their own patois, but also spoke good French, and I would be surprised, quite illogically, to receive a gentle 'Bonjour, monsieur' from an old woman riding through a cane-field on a donkey, seated sideways in billowing skirts and tiers of petticoats, her headdress tied in a careful arrangement of points.

My education in food and drink covered a wide range, too. Every meal started with a few glasses of the ubiquitous and potent 'poonch', either 'blanc' or 'vieux'. It was a small glass of rum, syrup of sugar, and a twist of lime. The rum and limes were free, and only the syrup had to be bought. We had meals at Les Pitons, at village inns, Chez Etienne, at hotels, clubs, cottages and houses. We ate hot crab and peppered steak, bananas in cream and bananas flaming in brandy. We ate pot-au-feu, papaya, and sea urchins, custard des crevettes, lobsters, eels, dorades and tuna. We sampled terrines and cheeses of France, and once we had cold rice pudding. Even that was good.

All these diversions ashore were fun, but my real reason for being in Martinique was to have work done on my boat in Grant's boatyard. On hearing Monsieur Grant's name, and on first seeing him, I had a feeling of relief that I would at last be able to talk in English without the effort of turning everything into French. It was a false hope, since M. Grant's last connection with England or Scotland had been through his grandfather.

The work went well. The yard was a small one, in a little creek at the end of a shallow bay, into which the foreman, Gabriel, had piloted me from a dugout canoe. On each side of the slipway there was shade from the trees whose branches nearly met overhead. Blackbirds sang in the branches and gave a happy start to each morning.

M. Grant's small schooner, *Le Merle Blanc*, lay at the entrance to the slip. As a young man, working in his father's timber-yard, he had started to build this schooner, with Gabriel as an unskilled boy to hand him his tools. As time went on, his friends asked him to do repairs on their boats and soon he was doing more work on other boats than on his own. He found that he needed more help, and in the end, a complete yard grew out of the small start.

They did good work for me. We checked the rudder stock and the pintles, drew the propeller shaft and found it all right, installed another Baby Blake W.C., and built a strong support for the little gas refrigerator. We added wooden flaps to the sides of the forehatch so that it could be kept half open when sailing. Then we painted *Leiona*, above and below the waterline. We launched her one afternoon, and I sailed off.

The last part of my trial sail in the Caribbean was to come, and it was this part that brought the first trouble. Light winds took me clear of Martinique during the night and along the lee of Dominica the next day. *Leiona* was steering herself well, sometimes with the wheel lashed, but often with it free, balancing easily with the steady beam wind. I wanted to visit the little islands of The Saints, 40 miles away, which sounded attractive, but I had been warned not to try entering after dark, as there were no guiding lights and the entrance was narrow and twisting. There seemed to be plenty of time. I came into the open sea north of Dominica in the early afternoon, with only 15 miles to go. Here there was a pleasant Force 4 wind, almost on the quarter, and we sailed at five knots with jib, staysail and mainsail. Five or six miles from land, the wind began to increase a little, and the waves became steeper, throwing *Leiona* about too much to let her steer herself. After an hour of this, I began to find the steering difficult. The boat was yawing and trying to come up into the wind. I knew that I had on too much sail, but I did not like the idea of trying to get it down by myself in that sea and wind. Although I had left it until too late, I had an idea that it would be all right for the next half-hour or so, and that by then we would be in the shelter of The Saints. I was wrong. The wind increased and so did the waves. It was an effort to stop *Leiona* turning into the wind, and I had to pull hard at the wheel, bracing myself as if at a tiller. There was a tremendous amount of noise from the wind, spray and waves. I felt frightened.

I thought I heard a slight noise which was not in the pattern of sound, something like a muffled crack. At the same time, the steering became much easier. Then the boat began to turn and to find her way into the wind. The wheel felt as if it was working, giving some resistance, but it had no effect on the steering. As *Leiona* came first beam-on and then nearly bow-on to the waves she rolled violently, and the spray drenched the sails and me. I managed to get the headsails off very quickly and then had a wrestle with the main.

The boom was flailing about and it was hard to gather in the sail and get some gaskets around it. The boom crutch was not of the scissors kind, which I think would have been impossible for me to set up, but it turned out to be not much better. It was like a large dinghy rowlock, stepped in a low bracket on the after side of the stern cabin, consisting of one long iron leg with a half-circle at the top. I put the leg into the bracket, but it was a job to get the boom into the small half-hoop.

After a struggle I got the boom into the crutch, sheeted it down tightly, and crawled aft to see what was wrong. I thought that the rudder must have broken, and at first the refraction in the water made me think it was swinging free from its pintles. When I saw that it was all right, I looked at the steering cables, but they were sound. It was then that I saw the cover over the steering quadrant had been raised and one of its planks had been forced upwards. I lifted the cover out of its fastenings, lashed it to the rail, and went back for another look.

The steering quadrant had broken. On one side the rivet clamping the inner and outer arms on to the rudder post had sheered, and the quadrant was moving loosely, not turning the rudder. The weight that I had felt through the wheel was only that of the quadrant plate and the cables. The boat was now lying across the waves and a great deal of spray was coming on board. I got out a big pipe wrench and tried to make a temporary tiller out of it, but it would not stay firm, and in any case, the leverage was too small.

Two things that I had stowed right in the bows, underneath the spare sails and warps, were the sea anchor and the iron tiller for the twin staysail self-steering gear, as I had foreseen no reason to use them between Martinique and Antigua. I got them out now to be ready for use, cluttering the forecabin and the saloon with the piles of gear which had been on top of them. The westward drift was fast, and I thought of stopping it before starting to work. The sea anchor had never been used, and I had idly left it in a tangled state. After one look at it, I changed my plan and decided to go straight on with mending the steering gear.

I took the bolts off the tiller and tried to clamp the quadrant arms together. This was impossible because the arms were sprung too far apart and the bolt holes would not line up truly. I tried various spanners and pliers, but nothing was big or strong enough.

It was not easy to work. The movement was very quick, and the stern deck was being swept by waves. It was hard to keep myself on board and to use tools. I could use only one tool at a time and had to go to the cockpit whenever I wanted to change tools. For the first time, I felt that I would have liked another person on board. I was feeling tired, frightened, wet and incompetent. I kept talking aloud to myself in an odd way, swearing in English and giving myself technical instructions in French.

I was getting nowhere until I suddenly remembered that somewhere on board was a bench vice. Luckily, my first search found it, and it did the job. It was strong enough to clamp the rudder-quadrant arms firmly, and I could steer again with the wheel.

I got the jib up and went down-wind for The Saints. A lot of time had gone by and the sun was almost setting. I hoped that we would get there before dark, but the chance was slight. I started the engine to add some speed, and studied and memorized the chart. The south-west entrance was the quickest to use, but it was not an easy one. It was narrow with a sharp bend, and beyond this bend was a lump of coral of uncertain extent. The coral was marked by a buoy, but the buoy was not lit.

After sunset, I reached the entrance and turned in with the last of the light. Before dark, I had a look from the cockpit at the steering quadrant. Something was wrong. I looked again and saw that the quadrant was turning with the wheel, but slipping slightly around the post, turning the rudder, but not at quite the same speed as the arms. I could think of three things to do: go out to sea and clamp it again, anchor then and there, or carry on. To start work on the gear would mean the loss of what little light was left. To anchor was not safe as the shelter was not good and the rocks were too close. I thought that if I went on slowly, steering very gently, the gear might last out long enough. It was a risk, but it worked.

I crept in as slowly as I dared and turned to starboard when I thought I could just see the unlit buoy. I held closely to my compass course. Some lights from the village ahead were a help, and I could see the tops of the hills beside me against the sky. What I could not see was how far the lower spurs of the slopes came out into the bay. In the darkness I saw close on my starboard bow two big, flat-topped objects that looked like coal barges and I passed near to them. In the light of a torch I saw that they were rocks. A little way beyond them I could see the sand of a small bay, and I anchored in three fathoms of water. I was very tired. It took me an hour to clear up the boat, which was in a mess with everything piled into the saloon and sea anchor, tiller, tools, books and cushions all flung about. I scrambled a bowl of eggs and fell asleep.

The morning light showed a pretty little bay enclosed by islands. Ahead was the main island, called Terre d'en Haut, with a sloping green hill, a small village, gaily-coloured fishing boats on the beach and an old fort. Astern were the headlands, rocks and coral which we had somehow cleared in the dark. It was not a happy memory and I wanted to get away from it. I sailed out of the wide north-western passage and along the coast of Guadeloupe in a light wind, on a bright sea with porpoises playing under the bow. That evening brought me to the small, round bay of Des Hayes, where nets and fish traps dotted the water. It was in the north of Guadeloupe, and was my last stop before Antigua.

My month of sailing in the Caribbean was over. It had been a good month, with only one real fright. At midnight I was awakened by peals of bells from the village. It was Easter Sunday and the first day of April. It seemed as good a time as any to go. I got up the jib, ghosted out of the bay and sailed 45 miles to Antigua, making for English Harbour.

CHAPTER THREE

Nelson's Dockyard

English Harbour was a good place for my month of preparation. Most of the dozen or more yachts usually there had crossed the Atlantic, and some had sailed much further. The advice of their owners was of great help to me, not only in its detail, but because it made an Atlantic crossing seem a normal and almost casual thing to do. All sorts of boats had sailed across, and all sorts of people, ranging from the huge schooner *Te Vega* with her crew of Swedish professionals, to small *Westering*, a converted lifeboat whose family crew included a girl three years old, her one-year-old brother and a dachshund.

It is impossible to think of English Harbour without thinking of Lord Nelson. Nelson's Dockyard is a lasting memorial to him, but nowadays the name most often heard and the influence most strongly felt there is that of Nicholson. Vernon Nicholson is best known as 'The Commander', and in English Harbour this lower but more exclusive rank enjoys greater prestige than the higher one of 'Captain', which is applied to every owner or skipper.

After the Second World War, Vernon Nicholson sailed into English Harbour with his wife and two sons, having crossed the Atlantic on their way to the South Pacific or around the world or to some alternative to the grey life of England or Ireland. They meant to stay for a while in Nelson's Dockyard, which was then deserted and almost unknown. Then they became involved in doing and arranging yacht charters, and in a few years found that they had become a thriving agency for charter and travel.

The influence of the family was widespread. When Vernon was not walking around the dockyard sorting out complications of charters and the other countless problems that boats raise, his voice was heard all through the Antilles during his wireless broadcasts. These were held twice a day and were conversations with the many boats sailing on charter under his agency. Emmie, his wife, had become the chatelaine of the Dockyard. She was the one to ask for domestic advice, and she was the unofficial but effective guardian of the happy and musical horde of coloured girls who flocked in through the gates each morning to do the washing and cleaning.

Both the sons had married Americans. Desmond and Lisa lived at the Dockyard where they helped Vernon and were in charge of the ship chandler's shop set up in the old Paymaster's Office. Rodney and Julie ran the office end of the business and the travel agency in St John's, the capital of Antigua, ten miles away.

It was hard to enter into any sort of activity in English Harbour without finding that some or all of the Nicholsons played a big part in it.

The whole setting was perfect for what I wanted to do, and the sense of history and romance was part of the perfection. Here was a place which had been the dockyard for the ships of Nelson and his captains nearly two hundred years ago, and it had changed very little. Nothing new had been built, and even the restoration of the roofs and timbers was not complete. Some of the old lofts and stores were just bare stone walls and pillars. The old capstans for hauling down the ships onto their sides for careening and scrubbing were still there. The Paymaster's Office had become a small shop for yacht gear, kept by the ever-cheerful girl, Ethylene, under Desmond Nicholson's guidance. The old Officers' Quarters were fittingly used as quarters for the captains of the yachts which made English Harbour their base for sailing and chartering. There were no roads, shop windows, neon lights or advertisement signs. There was no electricity and only one telephone.

It was easy to picture what it must have been, with the crowded masts and square yards of the frigates and ships-of-the-line along the dock or moored in the bays of the small land-locked harbour. The noise and bustle of hundreds of men moving between the ships and the buildings, carrying casks of water, food and good West Indian rum, and hauling sails, ropes, shot and bags of gunpowder from the stores, had now changed to a sleepy quiet. In the mornings, the stillness was broken by the arrival of an occasional taxi, the sound of a yacht charging its engine, and the singing of the girls washing clothes beside the capstans. The hot afternoons brought siesta time except for those who had to work. In the evenings the only sounds were voices and laughter from a party in one of the yachts.

It was not always quiet, even in the afternoon. Sometimes a fleet of taxis would pour into the Dockyard, filled with visiting tourists from a hotel or cruise ship. Someone had told me of these invasions, and had said:

'You will see the most amazing sights. It is really extraordinary to stand here and watch them, wearing the most fantastic hats and clothes.'

He was right about the hats and clothes, but I was not sure that the tourists did not have stranger sights to look at. They, at least, had a sort of uniformity in their striking forms of dress. The residents of the Dockyard were apt to produce a far more startling variety. Living in boats often leads to an expression of individuality in dress and appearance, and hot climates give a wide scope for such expression. English Harbour had a good share of this.

In the afternoon a visitor to the Dockyard might find a kaleidoscopic variety of sights. There was a squat little blue ketch looking like a nursery houseboat, with a small girl clutching her even smaller brother, saying they had sailed across the Atlantic and were on their way to Australia. Beyond the ketch was an old white yawl, with an enthusiastic American woman on board, her hair a bright orange, giving voluble and unnautical orders about 'those ropes and things that pull the sails up and down the mast'.

The visitor might be overtaken by a rosy and cheerful Englishman in white shorts being towed around the Dockyard by a Boxer dog, while

Emmie Nicholson floated by in a cloud of small dogs and cats. He might be impressed by the monocle of the debonair captain of a majestic black schooner, or startled by my Malay sarong and Bond Street hat. Maybe a tall Irish engineer, with long hair and a square black beard, would clump by in his heavy boots, carrying some oily part of a dismantled engine. Or a slim blonde would float up to a tall masthead in a bosun's chair, dressed for sandpapering in an almost invisible bikini.

These odd sights in English Harbour were nearly always provided by the white people there. The coloured folk were far more conservative in their dress.

I had written to Vernon Nicholson, asking if I could come there for my preparations and could call on them for help and advice. I would have found it very difficult without their many kindnesses. All the family helped me. Rodney and Julie had made just the right arrangements with the 'banana boat' shipping line to bring me and my gear from England to St Lucia; Desmond got charts and a wireless transmitter for me; Vernon found two good men to help work on my boat and gave me into the charge of his shipwright, and Emmie and Lisa gave me help and hints on my housekeeping. Most important, they all gave me confidence and encouragement, and with their experienced advice, relieved many of the fears I had felt about sailing the Atlantic.

It was noon on Easter Sunday when *Leiona* reached English Harbour, and no work could be started for two days. In those two days my crew problems began to sort themselves out. A letter had arrived from Michel Peissel, saying that he and Marie-Claire were in New York and ready to come when I wanted them. There was also a letter from an American cousin in Pittsburgh saying that a young man in his office, who had been in the United States Navy, was eager to come. If I wanted, he could bring a girl to cook. I wrote him to find out what he knew about sailing, as I was not sure of the capabilities of the two French, nor entirely happy about my own, and I wanted to make certain that anyone else who sailed with us was at least used to small boats.

The next morning a very nice German, Detlef von Eiberg, who had commanded a submarine in the war and was now skipper of *Harebell*, came over and said:

'At a party last night I found someone who would like to sail with you.'

I asked who he was.

'It is not a he. It's a she; the one you see over there in *Pas-de-Loup*.'

'Detlef, for heaven's sake, that idea is of no use at all. I want one more crew for the trip, but I don't think a seventeen-year-old blonde, even a pretty one, is the answer!'

'That's where you're wrong. She is over seventeen, a very good sailor, and has just brought old *Dayspring* down from Miami, with only two coloured hands, through lots of troubles. I spoke to her last night about your plan, and I think that she would like to go with you because she wants to get back to England. You should have a word with her.'

I walked across to *Pas-de-Loup* and spoke to the girl. She was Murlo Guthrie, John Guthrie's sister, and was five years older than her teenage appearance. Detlef was right. She was a very good sailor. In the past two years she had crossed the Atlantic twice in the big *Te Vega*, and, with her brother and two others, had brought a new boat the size of mine across the Gulf of Mexico.

Murlo had not decided whether to take a job in Antigua or go back to her mother in England. She wanted a few days to think about it, as she had only just got in from the Miami sail, which had been a hard one. I told her to let me know soon, and said my boat seemed sound enough, the weakest thing about her being myself. A few days later she told me she would come, and also said there was a young New Zealander, aged twenty-one, aboard her brother's boat, who also wanted to go to England, and would be able to get away at the end of the month.

This was more than I had planned for. There was room for only four in the two cabins, but, of course, someone could sleep in the saloon if necessary. The real problem was water. *Leiona* carried 100 gallons in one big tank, and another 10 gallons in two plastic bottles. I had thought that the 4,000-mile journey would take us about six weeks if things went well, but that we must allow for eight or even ten weeks in case of trouble. This problem could be eased by stopping for water in the Azores and maybe Bermuda.

Planning on half a gallon a day for each person, our water capacity was more than enough for three people, all right for four, but too little for five. I talked to Murlo about this shortage if we took five, but she said that Peter Vandersloot would be well worth taking. He had a lot of sailing experience, was a deisel engineer and was always cheerful and confident. A way was found around the water storage problem. I bought six more big plastic water bottles, holding an extra 36 gallons, which we lashed under the dinghy on the cabin top. So Peter came as well. Whenever anything went wrong he was confident it would come right, and usually managed to make it so by his skill.

Before Murlo decided to come, five others came to see me and four more wrote. The difficulty I found was not in getting a crew, but in choosing the best and giving polite refusals to the others, including the Pittsburgh American.

April was a busy month. The only fault I could find with English Harbour was the difficulty of getting things; if something was needed which Desmond's little shop could not supply, a search in St John's was necessary. This involved a ten-mile journey each way by taxi and took up a lot of time. It was usual to share the taxi with others, and this took still more time, especially as the rendezvous would, for convenience, be at a hotel or bar, which naturally led to rum punches and longer waiting. Apart from these long and often fruitless searches, the work on the boat went well.

Paddy Adams spared me what time he could between the urgent tasks of repairing the generators and plumbing systems of charter schooners which always seemed to go wrong just as they were about to start on a new trip. He checked over all the pumps and winches, and we worked out a new arrangement for the rudder-post quadrant. A square-sectioned socket was welded and attached to the quadrant as an extension of the rudder-post. Now the tiller could just be slipped over the post and secured by a pin, without the complications of several bolts. A full-size pipe-cot was acquired and cut down to fit in the forepeak. No one ever slept in it, and it would have been a tight fit, but it was useful as a shelf for warps and sheets.

My two steady helpers, who worked only for me for three weeks, were Mr Potter and Rupert. In the Caribbean there seemed to be a very delicate important distinction in the form of address. I never discovered how it was defined, but I know that if I had called Mr Potter 'George', or even 'Potter', our relationship would have been strained. And I think that Rupert would have been surprised and embarrassed if called 'Mr Henshaw'. They were as punctilious about this etiquette with each other as with me. I wonder if by now Rupert has been promoted to surname status. They were good workers.

They arrived on time every morning, each with a lunch-pack and a bottle of fresh lime squash. They discovered the small gas refrigerator and politely asked to use it. The two bottles soon became four, and after a while they extended the privilege to their closer friends. My cooling space became too full for my own things, but it was not for long, and it made my crew happy.

We laid out the sails and ropes in the handy open spaces on the dock, and went over them all, sewing, patching, whipping and splicing. We worked down from the masthead, going over all the rigging and opening all the parcellings to see if the splices were sound, which they were. We made a rope ladder with lovely greenheart rungs, to be able to get up the mast without using the bosun's chair. It seemed a good idea at the time, but did not prove to be a practical one.

We fitted grommets and splices to the awnings, made by Mr Potter's wife, and took out of the boat and piled on the dock every piece of gear, clothing and equipment that could be moved. The pile looked bigger than *Leiona*, but by careful stowing, and a certain amount of hard pushing, it all got back again, with more as well. We varnished or painted the new fittings as they were made, and finally covered the deck with non-slip paint.

Mr Gilbert was one of the select users of the refrigerator. He was the Dockyard shipwright, employed by Vernon Nicholson. Vernon was the only person who called him 'Gilbert'; no one else even thought of doing so. He was an expert and a philosopher, and it was a joy to watch him work. He never seemed to be in a hurry; everything was easy and everything fitted for the first time. He would not be rushed into doing anything. He would think in silence and then say:

'Now let us examine the problem and discuss it fully. Then we can best see how the difficulty should be approached and try to find the right solution.'

All his solutions for my difficulties were good.

The cockpit was too deep and the floor did not give proper foot support when the boat heeled, and was too low for comfort even when level. We devised a narrow raised grating to fit athwartship across the middle of the cockpit, easy to remove, but affording a better foot-rest. It had to be in two levels to allow the locker doors to open. The finished article was beautiful and became indispensable.

A small detachable steering seat was put in which gave the helmsman a better view, especially when the dinghy was stowed on board. A mahogany rim was put around the wheel. I have never liked the look of it, but it has added to the comfort of steering and removed the danger of having a foot or hand caught in a spoke if the wheel should spin.

The biggest work was a standing boom-gallows. I had always thought that they were good things to have, and after wrestling with the boom and the crutch near The Saints I was determined to fit one. I would hate to be without it now. It is not a very elegant fitting, rather like a football goalpost, but it is strongly made of greenheart, and does its work well.

Mr Gilbert went over all the problems, and drew out the plans first. Then we spent a morning in St John's choosing the timbers and having them cut to working sizes. The most awkward piece of work was making a knife-drawer. Murlo found an empty space near the galley sink and wanted it filled with a drawer for the cook's knives and oddments. One was made, of a strange design and shape because space had to be left in it for the water-pump handle. Now it is hard to believe that it has not always been there.

The day came when there was no more work on *Leiona* for Paddy, Mr Gilbert, Mr Potter or Rupert, and I had some time to think. There was still a week of April left. Michel and Marie-Claire were due to arrive on the 29th. Murlo was away sailing to Martinique from where she would bring back gas bottles, wine and cheese. Peter was tidying things up in *Pas-de-Loup*, but was free for the day, so we went out for a sail to see if the new gadgets worked. Bruno Brown, from *Freelance*, came along with us and was pleased that we used sail all the time, even for coming back to the dock. Ian Spencer brought out his schooner, *Zambesi*, and took some good photographs of us. Bruno did a surprise test of my man-overboard drill by leaping into the sea. It was a long swim back, so we relented and picked him up. All the new fittings seemed right.

For the next few days I had the boat to myself which gave me a chance to do some letter-writing and to get the charts and sailing directions sorted out. There was still one thing that I had not done and it was too late to do it now. One of my aims in my March sailing had been to practise my sextant work. In fact, I found I was either busy sailing, or there was something else to do, or there was land below the sun which left me no horizon to use. My lack of practice and of confidence may have had some effect on the route I decided to take.

I planned to go generally northward from Antigua, keeping as far to the east as the winds would allow, until we picked up the Westerlies. This would probably mean leaving Bermuda several hundred miles to our west, and making the Azores our first stop.

The sailing distances would be roughly a thousand miles to the region of Bermuda, two thousand from there to the Azores, and another thousand to Gibraltar. Going to Bermuda would not add much to the distance, and it would help in replenishing water and fresh food, but I did not want to go there, for three reasons. One was because it was an expensive place and we would be bound to spend a few days and a lot of money there.

Another reason was a fear that some or all of my crew might find a thousand miles of sailing quite enough and want to stop, or I might even have the same idea myself. If we got as far as the Azores for our first stop, I felt that we would probably all feel like completing the comparatively short distance left.

But my main reason was the thought that my navigation might not get us to Bermuda at all. The island was small, very low, with reefs around it, and I might not find it. There seemed to be much more chance of finding the Azores, good high islands, spanning four hundred miles, and I would have more time to perfect my navigation. Edward Allcard had agreed that the route leaving out Bermuda was a good one, and both Murlo and Peter were happy about it, so the plan was adopted.

The final stage of the preparations was to load water, food and fuel. We had to get them first. The water and fuel were easy enough, except for the awkward deck stowing of the extra water cans. When we had finished, we had on board 140 gallons of water, 140 gallons of diesel fuel, eight of petrol, four of paraffin, and about three gallons of assorted oils and grease.

Food was a much greater problem. We had decided to have ten weeks of good scale living and a further two weeks of survival scale. For five people, this meant a lot of food, and I was appalled at the quantities. Murlo, who had been in charge of stores and feeding on *Te Vega*, worked it all out and discussed it with me, and then we went shopping. Most of our stores were in tins. Dried soups and vegetables would have taken up less room, but they needed extra water. It seemed easier to have the water already mixed in the tins, and so increase our liquid cargo. There was room to stow them in the lockers below the bunks. We wanted to have enough variety in our meals to give interest to them, but we also wanted to keep our stock of foods simple.

The finding, buying and stowing of stores was easier if we bought large quantities of a few items. We kept the selection down to three kinds of soups, two kinds of cold meats, tinned stew, steak and sausages, a few varieties of tinned vegetables—especially tomatoes—tinned and dried milk, tea, coffee and cocoa; tinned butter and margarine, cheese, cakes and puddings, chocolate, and enough sauces and flavourings to ring the changes in the tastes. Then we bought what seemed astronomical quantities of flour and sugar, fifty pounds of rice, fifty pounds of onions, seventy pounds of potatoes, two huge stems of green bananas and a pile of small, hard cabbages. We took some fresh meat, fruit and soft vegetables for the first few days, and big nets of limes and oranges. Our liquor stores were rum, whisky, gin and lime juice, with no room for beer, much to Peter's sorrow.

The last delivery was a case of thirty dozen eggs, over which we had a lot of discussion. Various ways of preserving them were suggested, and in the end, we chose the simple but laborious method of boiling each egg for seven seconds and turning the boxes over every week. The method worked, and we were still eating them seven weeks later.

Peter had another shock to add to his disappointment over the beer. His main memory of his last long voyage was an almost unbroken diet of baked beans. To his dismay, the first crates he saw being loaded were the hated beans and, worse still, a favourite of mine, tinned rice pudding. He became happier when he saw a case of New Zealand mutton stew going in as well.

As usual in the Caribbean, we were met by a crisis at the last minute or at least, on the last day. Our staple foods for starch and bulk were to be rice and potatoes. When Murlo went to buy them, she found there was a famine of both in Antigua that week. The ship which brought all the island's rice every two weeks had gone in for repairs and no substitute had been arranged. By the time we started our search all the rice had gone from the shops, and there had also been a rush on potatoes.

We were in a fix. These two simple items formed an important part of our stores and we could not happily sail without them. It was infuriating to be delayed by something which could not have been foreseen, and we foraged around to see what we could find. At the cost of a lot of time and taxi money, we managed to collect two or three pounds at a time from many scattered shops. By asking everyone we met, and sometimes being told that 'Maybe the sister of my uncle's wife would have a handful or two of rice to spare', we managed to collect more.

Te Vega's stores produced some tins of potato flakes which were of especial use in bad weather. Ian and Cynthia Major did us a great kindness by presenting us with boxes of the best quality packaged rice. With these gifts and with our taxi-searches, we were at last fully stocked. The pile on the dockside was enormous. In all, with fuel, water, gas bottles, food stores, and ourselves, I reckoned that we were giving *Leiona* an extra three tons to carry. It was our last afternoon. There was no more to be done except to load the stores and lash everything down. My crew knew where to put things and how to stow them, so I left the boat alone and went off to a deserted beach.

The trials and preparations were over. In half a day we would be heading for the Azores, across three thousand miles of ocean. I wondered how others had felt before their first voyage. It seemed a long way to go, and we would have many weeks on our own, with no one else to help us if things went wrong. I did not see why things should go wrong, as we had taken a lot of trouble over getting them right, but it was a sobering thought. The boat and gear seemed sound and we had plenty of food and water, unless we were badly slowed down by some accident. We had medical stores to deal with small illnesses and injuries, and we hoped not to meet major ones. John Slominski had given me some powerful pills to deaden the pain in case of appendicitis. I thought that our two worst enemies could be bad weather and bad tempers, and I hoped that we would be able to avoid them or deal with them.

CHAPTER FOUR

Hunting the West Winds

We had a perfect day for our start. Early in the morning the wind was blowing hard and outside the harbour the seas ran strongly, but as the sun climbed up the wind eased. By noon it was just right for full sail and the sea was specked with white horses, or as Mr Potter said, 'the sheeps sure was out to graze.' News of our leaving had got around the Dockyard, and as we made ready to go, people began to gather on the quay beside us.

We wanted to get away under sail. This meant tacking out of the narrow and winding channel. With all the stores we had put in, *Leiona* was heavy and low in the water, and I thought it a safe thing to have the engine running in case we needed it at short notice. When we were ready to go I pressed the starter. Nothing happened. My young crew had been entertaining their friends on board late into the night, and the batteries were flat. I thought that we had remembered everything, but Peter and I had each imagined the other was checking the batteries. There was not even enough power to turn the little generator which would have started the big engine. It was now time to go and I did not want to hang around any more, or to begin playing about with starting cords, so we sailed away in silence as a boat should.

In sailing, as in war, I have found that big and distant dangers are often forgotten in the concentration on some small but immediate difficulty. Certainly, on that morning of Wednesday, the 3rd of May, I was more concerned with not getting *Leiona* stuck on a shallow in English Harbour than with the problems which four thousand miles of sea might produce. Peter had sailed with me once for an hour, but the others were having their first trial of *Leiona*, and of me. We worked our way slowly out of the harbour in short tacks, under critical eyes from the shore. There was one muddle over the hanks of the staysail, which Peter and Murlo put right almost before it happened, and then we sailed on. Clear of the two headlands, we eased the sheets and ran off to the west, a steady wind pushing us along.

For five miles we sailed past the low hills on the south of the island, and then swung up to the north to use the shelter of Antigua and then of Barbuda before our long passage to the Azores. I wondered what the others were thinking about. They all seemed very happy and full of confidence in themselves, in the boat, and, much more surprisingly, in me. It did not seem to worry them that they were setting off on a very long and lonely journey, relying on me to find a group of small islands as my first attempt at navigation. Maybe they were comforted by the thought that the long coasts of Europe and Africa lay beyond if we did miss the islands.

I had plenty to think about. I was not really sorry to leave Antigua. It had not brought as much interest or amusement as had Grenada and Martinique. English Harbour had been great fun and left many happy memories, but it was an isolated piece of boat-world and would have been fun anywhere. My time had been occupied in getting ready for this journey, and I had got to know little about Antigua. Even if I had, it did not seem to have the attractions of the other islands, either of people or places. Unlike the others, it received the big jet aircraft which brought with them the boom of tourist trade. Hotels, shops and restaurants were big, brassy and expensive, and everything seemed geared to the almighty dollar. Nor did it have the tropical beauty of the islands to the south. By comparison, it was flat and dry, with shabby villages made of crude timber, corrugated iron and packing cases.

The cane fields gave some green relief to the bare earth where cotton, potatoes and tobacco grew thinly. Antigua did have places of beauty, but they were best found by looking outward from the island. The views from the tops of the ridges, looking down the wide valleys with their tall spikes of yellow flower above the sisal plants, were wonderful. The waters in the jagged small bays, broken by rocks and coral reefs, gave fascinating contrasts of blue, green and purple to the yellow and brown of the land.

I would have liked to have spent more time in the Caribbean, which had given me a very happy month of sailing, and delightful weather, but I felt that if I did not sail to the Mediterranean now, all sorts of things would crop up, and I might never get started.

I wondered how things were going to turn out. So far, from the buying of *Leiona* to the finding of a crew, things had gone too smoothly to be true, and I wondered when the troubles would start.

From what I had heard and read of small boat voyages, most of the troubles were caused by the crew. It was asking a lot, even of the most easy-going people, to shut them up in a confined and uncomfortable space, with little privacy and with the added strain of monotony and anxiety. I wondered how my crew and I would all get on with each other. It seemed a gamble, and perhaps not a good one. Murlo and Peter had sailed together, and from the little I had seen of them, they got along all right. Michel and Marie-Claire presumably liked each other, but they were strangers to the rest of us, and I was a stranger to everyone. Peter and Murlo, in spite of their youth, had much more experience than I had, and I hoped they would not become irritated by my lesser knowledge.

All four were nearly twenty years younger than I was, and I could see that this gap in our ages might lead to differences in ideas, although it might be a help to me in exercising gentle control. I realized that the control ought to be gentle, but there would have to be some. This was probably going to be the hardest thing for me, as I was used to the giving and taking of orders, and to the acceptance of discipline while my young crew were not. I could see the danger of making them sour by being too strict, and I hoped that I would be able to ride them with a light enough rein to avoid unhappiness. In fact, for no apparent reason, and with no apparent effort, we had no quarrels from start to finish. Maybe one reason was that we did not become tired or overworked. The spread of duties was easy.

My job was to be captain and navigator, and to keep a watch as well. Peter was the engineer and electrician, and was able to navigate if anything should go wrong with me. Murlo was in charge of stores, and galley. Michel was bosun, and had the care of sails and cordage and their repairs. Marie-Claire was to help Murlo. All of us kept watches, except on cooking days. I was off the cook-list, as navigator and general stop-gap, but each of the others took a day in turn. The cook did all the cooking and washing-up for the day, but had no watch to keep and so got a full night in bed.

That left four of us to keep watches each day, with two three-hour watches and two nine-hour rest periods. Only one person was awake at a time, unless help was needed. There was a case for having two at a time on watch, letting them sort out their own share of rest, but I was against it as it only meant a waste of effort and a lot of talking and moving about, disturbing those who were sleeping. Normally, one watch-keeper was quite enough, with nothing else to do but steer. I hoped that the solitary person at the wheel would stay in the boat, and not go overboard—the biggest argument against having only one man alone on deck.

Perhaps three hours of steering on one's own was too much at night and it might have been better to have settled for two hours on watch and six off, or to have had three hours on by day and two by night. The advantage of having three hours on watch and nine hours for rest, among four watch-keepers, was that everyone's watch came up at the same time each day and night, making it easy to get used to a rhythm of steering, eating and sleeping.

These duties and timings did not always work out quite as exactly as planned, but they were a good basis. Murlo did far more than her care of the stores implied; she was not only the best at steering and sail-making, but became the neatest and quickest navigator. The simple navigation that

I used was so easy that I taught it to her and Michel in a day, and she liked to take and to work the sights. Michel took a liking to his care of ropes, and extended this to whipping and splicing every piece of line for which a use could be found. Soon, any bit of gear which could be lashed, hung or carried had some of Michel's handiwork attached to it.

Peter could do everything, and he did. He would take on the most difficult and unpleasant jobs without any hesitation, and with his banjo and his songs, he was our main source of entertainment. He also had a great capacity for sleep and freedom from worry. Unless called from his bunk, when he would appear at once, the strangest and most alarming noises on deck failed to disturb him. As planned, he was on watch from 6 to 9, morning and evening, and I took over from him. I wake early, and I was usually on deck soon after 6 a.m. I had seen how Peter hated getting up at what he thought was much too early an hour, so I changed morning watches with him. He used to go to bed about 9 p.m., but even after nearly twelve hours' sleep it was an effort for him to appear for breakfast at 8.30.

Marie-Claire steered and cooked well when she was able to work. Unfortunately, in anything more than a slight sea, she was laid low by seasickness, but she made quick recoveries, and was a charming companion, adding much to the happiness of the ship.

The ship's language had to be English. Michel was completely bilingual, and, in fact, preferred to use English for figures. Murlo and I could get along well enough in French, and Marie-Claire struggled in English, but Peter's only knowledge of French was some strikingly up-to-date slang. So in general conversation we spoke English, adapting it to Marie-Claire's ability, which increased rapidly. Some of her phrases were delightful and they are still used in *Leiona*, years later.

After our normal Anglo-Saxon understatements, it was a little unnerving to hear Marie-Claire ask:

'Keith, do you think that tonight there will be a tempest?'

But it was almost worth being awakened to be told:

'I think now we are nearly not going. The wind is very few.'

She was brave, because everything was strange and probably meaningless to her; she was sometimes sick and I am sure that she was often frightened. In spite of this, she managed to stay cheerful, and I think that there was only one time when she really wanted to be somewhere else. There were several times when I would gladly have changed places with someone on a safe piece of land, and I had only myself to blame for being in a boat.

I felt that our journey had not really started until we were clear of Barbuda and on our way to the Azores. We left the shelter of Antigua in the afternoon of the first day and sailed on northward in the bigger waves of the open channel. We could still see the island low astern at sunset, and then we were alone in the dark sea.

Barbuda and its shoals was out to starboard, five miles away. I had sailed there in *Freelance* for a day of swimming and fishing. It was a low, flat island, like a Pacific atoll with beautiful shallows and bays among the reefs. It had no lights, and I hoped that we would not come any closer to it than our five miles. The wind got stronger during the night, and we sailed along fast, in the sheltered water. By 3 a.m. I judged that we were clear of Barbuda, and by then we could feel the open sea.

We altered course a little to eastward and settled down for a month of sailing with only our own company. I had feelings of relief that we were on our way at last, and of excitement about what seemed to me the most difficult undertaking of my life. The young did not seem to have any feelings at all about this. The wind was good, the weather was warm, the boat was heeling over and hissing along, and everything was all right. The girls wore their bikinis and we made rum punches to celebrate our first day at sea. Peter's banjo came out and kept us singing. We settled into the easy rhythm of watch-keeping and everyone was happy.

Our happiness came to a quick stop in the evening. About an hour before sunset I was in the stern cabin, which I shared with Peter. Looking through one of the stern scuttles I thought that I could see the rigging-screw of the backstay moving slightly, as if it were slack. I went to have a look, and found that the stay was loose. I could see nothing wrong at the masthead, but when I tested the jibstay, I found a bit of slack there, too. Then I had a better look, from the beam, and saw what had happened.

The masthead fitting had broken. The iron extension which held the backstay had torn apart from the weld at the cap-band, and was hanging downwards, held on by a fraction of an inch of welding. This was a fitting about which I had been doubtful, but had done nothing to improve. I felt sick with depression. This standing backstay was the only support the mast had aft, as there were no running backstays.

We got all sail down and started work. The first thing to do was to secure the backstay at the masthead somehow, and then think about how to fix it properly. I said to the others, 'I am sorry to have brought you to sea in a boat which starts to break things on the first day.'

Peter lifted spirits by saying cheerfully, 'Hell, that doesn't matter. Boats are always breaking things.'

As usual, Peter took on the hardest job. We put him up the mast in a bosun's chair to fasten the stay and the broken extension by a strop of wire and bulldog clips. It was not easy to do, as the swing at the masthead was violent, and even with a down-haul from the chair it was hard to keep Peter steady. We took the topping-lift off the boom and set it up, doubled, as an additional backstay. Peter got his job done before dark, but was not very happy about the strain it would take.

We decided not to use the jib until we saw how well things went, and sailed under main and staysail, which was not a good balance. The jib was a big and powerful sail, and we lost a lot of speed without it. When the immediate repair had been done the problem was what we were going to do next. It was asking too much to disregard the break and carry on to the Azores. An obvious course was to go back to English Harbour for repairs, but it seemed an awful thing to do. Apart from sailing for a couple of days in the wrong direction, none of us liked the idea of turning back to our starting-place, and the effect on morale would have been bad.

There was a more practical reason for not going back there; work in Antigua was slow, especially when it involved welding and journeys to St John's. I could see that things would drag on and on, and that in the end our start would be late enough to come into midsummer and the risk of having the weather disturbed by hurricanes. The only other thing to do was to go to Bermuda. It was about 700 miles away, in more or less the right direction and not far off the course we had planned. We could be sure to get a good repair job done there, and we could fill up with water and fresh food.

I could only see two snags. One was that it would take us a week to get there and we might meet some bad weather, which would be no fun with a badly-stayed mast. The other was my old fear that I might not find Bermuda. These were private thoughts only, and did not seem to occur to the others. We altered course and headed for Bermuda.

Our troubles had not quite ended. Before we sailed Murlo had discovered that Marie-Claire's birthday would be on our third day at sea. We had hidden on board some wine for a birthday dinner, and there was to be a birthday cake made by Murlo and Crêpes Suzettes by Michel. The morning of the birthday brought some squalls, and the first gust of one of them led to the disappearance of the newly-mixed flour, milk, and eggs behind the stove and down the narrow gap which then surrounded it. That was the end of the birthday feast, and anyhow we soon had something else to think about.

The wind fell light in the afternoon, so we set the jib again. This was a mistake, because an hour later the strop at the masthead and what was left of the weld broke away. It probably would have been wiser to have cleared it right away at the beginning, but we had not been happy about relying entirely on the topping lift for a stay even when doubled, as it was only a light nylon line. As a result we were lucky to miss a tragedy by seconds. When the strop parted, under tension, the upper end of the stay and the heavy bar with its jagged broken face fell with a thud onto the stern cabin, just where Murlo had been sitting. If she had not moved away a few seconds earlier she might have been badly hurt or killed.

We tried a new system of rigging. The nylon topping lift had too much spring in it to give a firm pull against the forward strain of the jib, so we took down the jib. The staysail set from the upper crosstrees. There were no backstays there to keep the mast straight, and we could see it bending forward. To stop this we made use of the two wire stays on which the twin down-wind staysails were to be set. These were shackled to the forward side of the mast, at the upper crosstrees. The lower ends were not yet shackled to their eyebolts on the foredeck, but were lashed to the shrouds. To use the stays, we had to pass them over the crosstrees and bring the ends aft, hauling them tight onto the Genoa cleats near the cockpit. This was not an easy job. It was Michel's turn, so instead of a birthday cake our present to Marie-Claire was the agony of seeing her beloved being bruised and banged against a swinging mast. We rolled down the mainsail until the head was at the crosstrees, and sailed on under a much reduced rig.

That afternoon I had my last wireless talk with Vernon Nicholson, three hundred miles away. When a gap came in the crowded conversations he was having with his charter fleet, I cut in with a call. I thought that maybe we were too far for him to hear us. To my delight he answered.

'Hello, *Leiona*. This is English Harbour. Very good to hear you. Is everything all right? Over.'

'Hello, English Harbour. This is *Leiona*. We are having a bit of trouble. Something has broken at the masthead and we have no backstay now. We have rigged up a makeshift job and are heading for Bermuda. Over.'

'Sorry to hear that. It will probably turn out all right. Things are never as bad as they seem. Anything else? Over.'

'Send our mail to the Bermuda Yacht Club. We should be there in a week. Over.'

'All right. We will do that. Good-bye and good luck. Over.'

'Thank you. Nothing more. Out.'

Vernon sounded confident and unworried. I felt that it was much easier for him to be so unmoved, in his house overlooking the Dockyard, with his wireless set in front of him and a glass of rum in his hand, than for me and my crew, alone and hundreds of miles from anywhere.

We had done the best we could, or at least the best we knew, and it was no good worrying. With Peter's banjo and a happy chorus we sang

5. Diana, cook and artist
6. Cala Badella, Ibiza

7. Cala Santa Galdana, Menorca
8. Villacarlos Port Mahon, Menorca

about *The Keeper of the Eddystone Light*, *The John B's Sails*, *The Girl with Freckles*, and other songs which brought memories of English Harbour evenings.

With good weather, rum punch at noon and cocktails at sunset, plenty of sleep, three good meals a day, and amusing company we were having a very pleasant sail despite our difficulties. I was feeling happier about my sextant work. The calculations of our position each day agreed fairly well with the dead-reckoning positions, and I began to be more confident about finding Bermuda. From my sights I did find that we were being pushed more to the west than I had expected. Some of this drift was caused by our lee-way, but the main reason was the westerly current which was moving faster than the ten or twelve miles a day shown by the weather charts.

I did not want to arrive on the west side of Bermuda, which would mean having the wind and current against us for a final approach. The best plan seemed to be to keep well out to windward, to the east, until the last hundred miles or so, and then to be able to sail downwind to our goal. To do this the best course was due north. I get a childish joy in sailing what I call 'at right angles to the world', on either north–south or east–west courses. These courses give me a feeling of getting over the water more quickly, and they certainly do give quicker changes of clock time and temperature.

I had thought that there would be a lot of boredom in days at sea on a long passage. I don't think that any of us found this true. There was usually something to do or see or talk about, and when there was not, time to read or sleep was always welcome. Another of my good resolutions was unfulfilled, like the one to practise my navigation while sailing about alone in March. I had brought some books on Spanish and meant to put in some hard work at starting to learn the language. In fact, I found that my days were pleasantly filled and that there were seldom long periods for concentration on my task.

I used to wake at about 5.30 a.m., and would be on deck for a while before taking over steering, usually from Murlo. I might do some sail-

3

changing before starting my watch. At 8.30 I would wake the cook of the day, and my relief, who was Peter unless he was cooking, when Marie-Claire took over his watches. At 9 o'clock it was pleasant to go below for a leisurely breakfast. Usually everyone was there except for the helmsman, although sometimes Michel preferred to go on sleeping, and Marie-Claire was apt to miss the meal if the seas were anything more than calm.

Murlo had the luxury of breakfast in bed the whole way across the Atlantic. She slept in the saloon, and breakfast time was less than three hours after she came off watch. She got her fill of sleep by staying in bed, which was beside the dining table, and having a quick breakfast in her baby-doll nightdress before going back to sleep again.

The next thing was to take a morning sight and work out a position line. Then by 10.30 or 11 o'clock, I would have finished the few odd jobs which seemed to crop up, and I would have an hour of sleeping or reading or even going through a Spanish lesson. This last did not happen very often. Around noon it was time for a latitude sight, for rum punches and Peter's songs. Lunch was always a cold buffet meal and was cleared away about 2 o'clock. The afternoons were filled with some navigation lessons and position plotting, and some sleep before evening cocktails and a hot supper. I was on watch from 9.00 until midnight, when I would call Michel, stay with him for a while, get a time signal and then sleep until dawn. And so the days went by.

The sea produced a variety of things to interest us, either in itself or in its creatures. Even hundreds of miles from land, when we had left behind the frigate birds and tropic birds of the islands, there were still visitors, and we were seldom without a pair of shearwaters skimming the face of the waves with their wingtips, or a sudden flock of little black stormy petrels. All these seabirds spent so much time flying and so little in feeding or resting that it was hard to understand how they kept their strength.

We saw few fish except the ever-present flying fish, and only caught one, a long, thin ocean gar of a funny green tinge. We looked at it for a day and then threw it away. Porpoises were always a delight, and were good company in the night watches, when they would come close alongside, surprising the sleepy helmsman with the sudden sigh of their breathing. It was fun to sail through a fleet of Nautilus, the small Portuguese Men-of-War with their purple mainsails set, seeing them capsize and then right themselves as our bow-wave pushed them aside.

We came into the Sargasso Sea four days after leaving Antigua, and had trouble with its weed from then until near Bermuda. We saw nothing like the impenetrable miles of packed weed, crawling with sea monsters, as shown in old books and charts. The biggest patch we saw was about the size of a tennis court, and it seemed fairly solid, with what looked like little crabs moving on it. We certainly never met weed so dense that we were checked nor did we have to avoid any, but odd pieces did foul our log spinner and line. This happened so often and made the log so inaccurate that for several days we gave up using it and just estimated our speed and distance.

On two days of flat calm the colours of the sky and sea at sunset were amazing. There was not the slightest trace of wind. I did not know that the open sea could be so still. Except for the wave of our own movement, using the engine, the water was without even a ripple. It was so smooth and motionless that it seemed to be thick and heavy, as if it were a lake of molten lead. The only flaws on the mirror of the surface were the little sails of the Portuguese Men-of-War, shining like flecks of silver on the lead. In the evening the cloudless sunset made the sky a gentle pink and grey, and the leaden lake turned into mother-of-pearl.

Each night we could see a difference in the positions of the Pole Star and the Southern Cross. We were moving north about two degrees of latitude every 24 hours, and after two or three days the faint Pole Star began to show itself more clearly, higher in the sky and brighter, while the Southern Cross grew lower and fainter, and then sank out of sight for good.

As long as we held to our northward course I was not greatly worried about our position. The compass and the North Star agreed, with adjustment made for the magnetic variation, so we were going in the right direction. The log was not much use because of the weed which kept fouling it, but our estimates of speed agreed very closely with the distances between the positions I plotted, so my sights and calculations seemed to be fairly good. We were travelling along the 62° west meridian, hardly moving away from it at all, and this gave the feeling of being kept on our course by a nice safe tramline. Then the time came when we had to move off it.

When I reckoned that Bermuda was just over two hundred miles away, we altered course from north to north-north-west. I hated leaving my imaginary tramline and setting off into the blue. I knew that it was a ridiculous thought, but I couldn't help thinking it. With only the ocean around us there had been nothing to show that my efforts at navigating were correct, but nor had there been anything to prove them wrong. We had gone eight hundred miles since last seeing land seven days before, and now we had a very small fixed point to find, two hundred miles ahead. We turned towards it at 10 o'clock on the morning of the 10th of May.

I had been told that it was possible to find Bermuda easily by the many aeroplanes landing or leaving there. We kept a good lookout for them but saw none. At midnight on the 10th I thought I heard a faint signal on the wireless from what sounded like a direction finding station and might have been the letters TUK. I could not find this group in my books, but the only place that the signal could come from was Bermuda, and it seemed to come from dead ahead. The next morning the signal could not be heard. We were motoring on in the lightest of winds. There was a feeling of excited anticipation. For some reason I remember we had fresh hot scones for breakfast. Soon after breakfast a small cargo steamer passed astern of us, heading westward. If she were on the track shown on the chart we would be further away from Bermuda than we thought. Shortly after the steamer had passed we picked up the TUK signal again, still ahead, so we felt that we were going the right way even if the distance was wrong.

We hoped that during the night we would pick up the loom of one of the powerful lights of the island. If my figures were right, we could expect to come within range of the big Gibb's Hill light some time after 9 p.m.

As usual, nothing interfered with Peter's sleep after supper, but Michel was too excited to leave the deck.

I told Michel to look not only over the port bow where I expected the light to show, but all around as well. It would be a pity to miss seeing the light because we were on the wrong side of it. At 10 p.m., from halfway up the mast on the rope ladder he called out that he could see a glow of light, fine on the port bow, and it did not seem to be steady. In an hour we could tell that it was Gibb's Hill, and by midnight St David's light was also showing. I felt a wonderful relief that we had arrived, and that I could now be confident in my navigation for the rest of the voyage. Five miles off-shore we hove-to until dawn, moving slowly northward.

Early in the morning we sailed into the buoyed channel and then into Great Sound, to move through Two Rock Passage to Hamilton Harbour. I had some cleaning up to do as we passed the small islands. I had never had much success with electric razors, and the importance of saving fresh water at sea gave me a good excuse for growing a beard, which I had long wanted to do.

Beards became the fashion on board, but not a very successful one. Michel produced a growth of scruffy patches, while Peter showed only a few lonely hairs dotted about his chin. My beard was all right in area, density, and length, but disappointed me badly by appearing as a grizzled mixture of brown, black and white with much more white than I thought my 43 years deserved. Anyhow, whatever its colour, it had to come off. The only person I knew in Bermuda was the Governor. I knew that he sailed, and in the last Bermuda Race had been in a dismasted yacht, so he seemed a good person to ask for advice. He had known me earlier as a Sandhurst cadet and as a cavalry officer, and I thought that it might be too much for his Grenadier Guards tradition if I appeared as a sailor with a week's growth of piebald beard.

We anchored off the yacht club, later moving onto a big buoy, and I went ashore to get permission to stay there, clear with Customs and Health, collect the mail, and find out about getting the mast fixed. I telephoned Government House and talked with the A.D.C. to ask for his

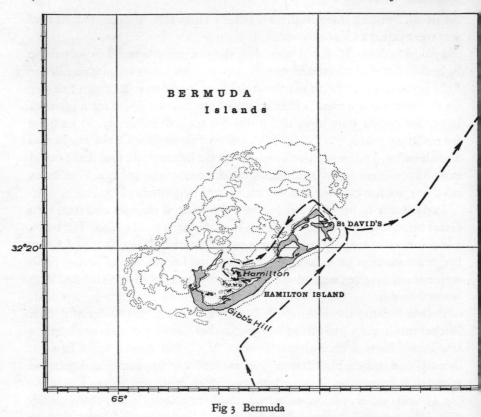

Fig 3 Bermuda

ideas. He told me to go to Bert Darrell. I suppose that anyone else in
Bermuda would have told me the same thing, and it was the best possible
advice. I doubt if anyone in the world could have done a better repair job,
and have been such a good host and companion while doing it.

Bert was a quiet man who did not talk much, but I learned a great deal
from him about boats and gear, sailing and racing, people ashore and
afloat, and the history of Bermuda, which was the history of his family
He was an expert craftsman and a perfectionist. I am keeping one of his
wire eye-splices as an example of perfect work, and yet he made it without

apparent concentration or effort, talking across the table as he sat behind the vice. I offered to do the parcelling and serving myself, but he said that none would be needed. The splice was left bare, and the strands of wire looked as if they had grown together without any ends to be seen or felt.

When his work was finished we all felt happier about the safety of the mast. Instead of the cap band, which left the end-fibres of the masthead exposed, he forged a strong helmet to take the strains of the rigging. He fixed a new arrangement for the mainsheet blocks and so removed the chafe which a bad lead had caused. He checked the rest of the rigging and spliced and fitted a pair of running backstays, drilling in two strong eye-bolts at the quarters.

As at English Harbour, I found that I was too busy about the boat to know and enjoy Bermuda properly. We were only there a week, and there was a lot to do. Michel and Marie-Claire managed to get around a bit, and did a tour of the island, both riding on one perilous-looking bicycle with an auxiliary motor. On two holiday occasions Bert took Murlo and Peter to crew for him, racing in his One Design sloop. They won both times, but I think it would have been a surprise if they, or rather Bert, had not. We had a bit of Bermuda's social life in the evenings, sometimes with Bert, sometimes alone or together as a crew, and once at Government House. I was able to make up to Marie-Claire for her lost birthday supper, and we all had a terrific dinner ashore which included Crêpes Suzettes. Quite unexpectedly Michel met his brother, Philippe, who had sailed from New York in his small cutter *Tethys*.

Sunny Bermuda belied its reputation by having more rain than sun while we were there, but even so it has left me memories of beauty. It was a lovely scene in the sun with clear bright waters, and white and pink houses in pretty gardens with green lawns. I wished that I could have seen more of it, but there was work to be done on the boat, stores and gear to buy and load, and we were eager to be on our way as soon as we could.

On our last day, the 19th of May, we motored across the harbour to fill with fuel and water, and had lunch ashore. In the afternoon we went back to Bert's dock to say good-bye. He gave us his good wishes for a pleasant sail to the Azores, and wanted us to let him know when we got there. Then he had to go. We finished the last jobs of lashing and stowing, and then sat about to chat and think. I find it best not to hurry a final action.

The wind was off the dock, and would take us well on our way. Two Americans with three children came strolling by. They were not the busy people that were usually seen in Bert's yard, but were more like the tourists of English Harbour. They were pleasant people and we found ourselves talking with them.

'Say, do you live here in Bermuda?'

'No. We have just been here for a week.'

'Where did you come from?'

'From Antigua, in the West Indies.'

'Gee, that's a long way to come. Where are you going next? Back to Antigua, or over to the States?'

'No. We are hoping to go to Spain.'

'Spain? All that way in that little boat? For heaven's sake. When are you going to start?'

'If you would just pass me that loop of rope, please, we will go now.'

'Right now? Well, for crying out loud! Fancy! Just like that.'

And just like that we set off on our journey once more.

CHAPTER FIVE

North Atlantic

Again we had perfect weather for our start. We sailed out of the buoyed channel and cleared St Catherine's Point and Mill's Beacon as the sun set. For the first time we were able to set our twin staysails. It was a wonderful feeling to take in all the fore-and-aft sails and to be heading to the north-east, the two staysail poles held still by their sheets and guys, and the twins drawing strongly and silently. The dark closed down over Bermuda, and that was the last land we saw for seventeen days.

For the next few days we had light winds, from between south and west, and we moved slowly north-east, very comfortably but doing less than 100 miles a day. We did not rig the self-steering gear to the rudder as the wind was too light to make it really effective and we had enough people to steer without any hardship. The twin staysails had been made for the strong winds of the Trades, and were too small for the light breezes we were having. We tried ways of setting more sail. The most effective combination was an odd-looking one with the Genoa to port and both the staysails to starboard, one in the normal way on its pole and the other set above it, flying from the masthead and sheeted to the end of the main boom. For want of a better name, we called this sail 'the flying raffee'. These three pulled us along well. There was no movement of the sheets and guys as they were made fast to fixed cleats and not to a moving tiller, and so there was no chafe. The drawback of the rig, and of having the wind astern, was the rolling. The motion was not particularly violent or rapid, but it was continuous and became irritating. Nothing stood still, and there was nowhere to lean nor any friendly lee bunk in which to put one's gear or oneself for safety. It was the only thing that came near to making Murlo bad-tempered, and it brought with it sea-sickness.

Seasickness affected us all in different ways. Michel was quite immune in any sort of weather, above or below deck. Marie-Claire was the opposite and became bunk-bound very quickly. Peter was almost unaffected, although he would begin to look a little green and lose some of his normal gaiety if things were really bad. He was actually sick only once, after drinking a tin of grapefruit juice. Murlo had told me before starting that she was always sick for two days after leaving port, and was all right after that. We left only three ports, but she was sick each time, even though the weather was not bad. Her sickness did not stop her working, and it was amazing to see her doing her turn as cook, apparently all right, but suddenly coming out of the galley to be sick over the side and then going back to the job again.

I have been seasick all my life, in large ships and in small boats. It does not upset me a great deal and does not stop my working, but I used not to be able to keep any food down. I tried all sorts of remedies without much success until I found one called Marzene. Now I always have some of this on board, and I take a pill before sailing in anything other than a flat calm, or whenever I think that the wind is going to get up. With its help I no longer have any worries about sickness unless I have to do a job which involves hanging head-downwards. Prolonged work with the sextant and navigation books is apt to make me a bit pensive. I had Marzene with me for this voyage, and I was only once caught out by the weather. Disappointingly, it was no help to either Murlo or Marie-Claire, but nor was anything else.

One thing that really proved its worth in the rolling, and at any time of movement, was the saloon table. It was a swinging table, weighted by 150 pounds of lead, and it became a very attractive resting-place for everything movable. It was remarkable how mugs, glasses and even bottles. would stand safely upright in really bad weather. We tried to hold to a rule of keeping it clear between meals, but it was not an easy rule to follow. It failed us only once, just before a meal had been cleared away, when it was covered to the full. A spare sextant was kept chocked on the saloon deck in one corner. One evening it moved and caught under the pendulum of the table at the end of its swing. As the boat rolled the other way, the table stuck fast, reached an almost perpendicular angle, and poured everything on to the floor.

Our 'bikini weather' sail from Bermuda was very pleasant, but we were becoming impatient and wanted to pick up the strong westerly winds to make a better speed to the Azores. On our fifth day out of Bermuda, when we were nearly five hundred miles to the north-east, things began to change, and very quickly. Before noon the wind veered to the north-west for a while, and then came back again to the south. We had changed sails to main and Genoa, but in the afternoon set our downwind rig again, with the flying raffee. The barometer was steady and the sky was clear. At sunset there was a little more wind, still from the south, and there seemed to be an increase in the feel of the sea. We took the flying raffee down before dark, in case the wind continued to grow stronger, as this sail was not easy to handle if the wind took charge.

I came on watch at 9 p.m. and soon had a sleeping boat to myself. For an hour everything was quiet and we rolled gently on our course, heading nearly north with a bit of east. Then the wind started to get up, and we were soon doing what I judged to be six knots, which is fast for *Leiona*. At the same time, some lightning began to flicker in the west, and a little later there was thunder. There was no change in the direction of the wind, but I saw that the barometer had started to fall from the steady 1020 millibars it had shown all day. We were sailing beautifully, and it seemed a pity not to use all of this good wind.

Before midnight we were really moving along, and the motion was becoming uncomfortable. The wind was still from the south, about Force 5, but in the west the thunder and lightning, in a thick mass of black cloud, was getting very near. I should have called Peter, sleeping only a few inches from me in the stern cabin, but I did not want to disturb

him, and Michel, in the fore cabin, was the next on watch. I lashed the
wheel and waited to see if the boat held a steady course. Then I went
below to call Michel, so that he could steer while I got the Genoa down.
I went through the saloon, swinging from the handgrips above my head.
The table was having a hard job to keep up with the rolls of the boat.
Murlo was asleep, held in her bunk like a cocoon by the bunk-canvas,
lashed up to the handgrips.

As I stepped into the fore cabin and put a hand on Michel's shoulder,
there was a shriek of wind. The boat heeled right over to starboard, and
stayed heeled, with the water running along the lee deck, piling up against
the coachroof scuttles. I whipped back through the saloon, calling out,
'Quick, Michel!', for which there was probably no need. I saw Murlo's
startled face pop over the bunk-canvas above me on the weather side,
and arrived in the cockpit just as Peter appeared from the other end. I
grabbed the wheel, slipped off the lashing, and turned downwind which
brought us upright. Peter said:

'Christ, that ought to have come down long ago!' He was looking
at the Genoa, which in the darkness seemed to have taken on an odd
shape.

Murlo was in the cockpit by then, and I gave her the wheel and a course
while Peter and I, followed by Michel, went forward to deal with the
Genoa. It had half dealt with itself. It had a wire halyard leading to a self-
stowing winch with a friction brake. The sudden pressure of the new
squall from the west had been too much for the brake, and the halyard
had run, letting the sail come half-way down. It was a lucky accident,
reducing the sail area more quickly than we could have done, and maybe
saving us from damage.

We got the big sail down and lashed it on the rail for the time being.
The pole had its own topping-lift and was well cocked up, so we left it
where it was until later, and thought about the next step. We were all a
bit startled and may have exaggerated the difficulties, but it seemed to be
a wind of a good Force 6 or 7, and maybe even more. We were moving
fast under the starboard running-staysail. This rig looked very unbalanced,

but, in fact, the steering was not hard. The staysail was a small one and was held so firmly, unable to flap or gybe, that I left it set.

I did not leave it for long. In less than half an hour there were some really big seas running, and the wind was even stronger. I thought we were going to get into trouble if we went on as fast as we were moving then, as we were beginning to pile up our own quarter-wave, which might have made the following seas break. It was the first time that I had been in bad weather in a small boat, and I thought that maybe I was making too much fuss about things, but we were not racing, so we slowed down. We set the working staysail, which had been left hanked to the forestay, and took in the solitary twin and both the poles. Our speed dropped and the motion was much easier, with less disturbance in the water. To keep the sail area as small as possible, the staysail was sheeted hard amidships with both sheets. Every now and again it would gybe with a bang, but it was not doing any harm, and I only hoped that it would not tear and blow out.

We were still moving at about four knots, and I would have liked to have gone even more slowly. I was afraid to take in the staysail and run under bare poles, because with the mast almost amidships, and the added windage of the dinghy above the cabin top, I thought that a combination of wind and wave might push the stern round, presenting the whole side of the boat to the waves. Peter and I talked about streaming some warps astern, but we thought things were not yet bad enough for that. We seemed to be going along without much water coming on board, and decided to wait a while and see how things went.

The barometer had begun to drop before midnight, and it went down a little more in the next two or three hours. Then it stayed steady at 1015 millibars. With the coming of daylight the weather seemed to grow worse, or maybe it was because we could see more of it. The wind held from the west and the waves marched on in long heaving ranks, the tall ridges being at times broken by an ugly-looking pyramid of greater height when two wave formations converged.

There were two swells, clearly marked; the lesser one from the south and a larger new one pushed on by the wind which had hit us from the west. While these swells were at right angles to each other they seemed to lead separate lives, but the one from the south was not big enough to worry about. Sometimes the line of the south swell would swing as if to join the bigger one, and then the two would either join forces, increasing the height of the big one, or would push against each other to form a steep peak which would tower up and then fall forward, overrunning itself in a white cascade, tumbling down the green slope streaked with foam. These cascades appeared to be four or five feet deep, double the volume of the crests which were breaking all along the tops of the waves. I did time the interval of the seas, but I did not write down the results and have completely forgotten them.

We were anxious not to exaggerate and turn a fresh breeze into a 'yachtsman's gale', but this one certainly seemed to be a real gale. We got out the *Yachting World Diary* and matched its descriptions of wind and sea with what was going on around us. Even with cautious estimates we reckoned the waves as not less than twenty feet high. The pyramids were much more than this, but luckily we never met one. We could see at least the characteristics of a Force 8 gale everywhere, and were inclined to think that maybe a strong gale, Force 9, was a better description. At times there were 'waves with long overhanging crests', but we could not agree if those waves qualified for the 'very high' description of a storm of Force 10. All this was academic, and we found the wind much stronger than we wanted, whatever its official designation might have been.

Our watches were cut down to an hour each, as steering had become a hard and uncomfortable job, with a lot of concentration needed to keep the stern to the waves. An odd thing was that this weather, the worst we had in the whole voyage, did not make Marie-Claire feel in the least sick. Even so, I did not let her steer, as it involved something more than just keeping a course, and I did not think that she could correct an error in time to stop broaching. It was lonely in the cockpit. The saloon doors had to be kept shut to keep out the spray and rain, and the noise of wind and wave covered any sound of life within the boat.

This was the hardest test the boat had had, and I wondered how she would take it, and if things were going to be worse. I felt frightened once, and opened a saloon door for a moment to look in, mainly, I think, for the comfort in seeing that there were still others on board, but also to see how they were taking it. I need not have worried. The saloon looked like a bar on New Year's Eve. All the young were there and awake, even Marie-Claire. They were on top of the world. The air was thick and warm with cigarette smoke, the table was covered with biscuits, chocolate, cake, mugs of cocoa and a bottle of rum. Peter had his banjo out and twanging, and they were all roaring with laughter and song. I shut the door and felt much better.

As had happened before, a minor irritation could push a greater danger into the background. One thing that made us all annoyed and uncomfortable was the poor protection our oilskins gave us. In England I had bought four new suits of what I was assured were the best possible oilskins. They were of proofed nylon, and I got each suit in a different and glamorous colour. The seller told me that they would keep out the heaviest concentration of any kind of water and that they would not become tacky, even in tropical heat. He was exactly wrong. They stuck together from the start, so that it was a job to put them on and they were useless against anything but the lightest drizzle. Luckily, there were some odd sets of old-fashioned foul-weather gear on board, and we managed to keep at least the helmsman dry.

Things got worse around 9 a.m. A black thunderstorm rolled up on us, with more wind and rain. It produced short but very violent gusts, which seemed to qualify for Force 10. It also brought sheets of almost solid, blinding rain, which was as cold and punishing as hail. But it had one good feature; the rain came with the worst gusts of wind and had a definite flattening effect on the crests of the waves, beating them into rounded masses, instead of the overhanging breakers which we were expecting from the increasing wind. The thunderstorm was the worst part of the gale. I was in my cabin, half asleep after handing over the steering to Murlo. Through all the noises I heard, very faintly, one sentence from her, talking to herself:

'I don't like this at all, not one little bit.'

I looked out to see what was the matter. I didn't like it at all, either. The cloud seemed right down on top of us, close around like a black circus tent. There was a constant roar and rumble of thunder, with frequent crashes right above us. Below the tent there was a weird half-light lit by flares and flickers of lightning with white and yellow streaks, forked and single, plunging down into the sea. It looked awful, and *Leiona*'s mast, the only vertical thing in sight and the nearest connection to the clouds, seemed to offer itself to the lightning. Nothing could be done and it seemed best not to think about it, so I left Murlo to her dislikes and went to sleep.

Steering was a strain, but it was also exhilarating. I found that it brought a thrill, sitting low in the cockpit, against the port side, with a safety belt hooked onto a stanchion, looking forward to check the compass course and aft to watch the waves. They looked enormous, their height seemingly increased by the angle of the boat as the passing wave lowered us into the bottom of the trough. That was the worst time, when the oncoming wave astern of us stood high above the boat, its crest curving and beginning to topple over. Then the stern would lift and we would be raised by the forward and upward swing of the advancing wave, and slide down its face, overtaken by the hissing, tumbling, frothing water of the breaking crest. This green and white turmoil was not dense enough to carry and raise the boat, and we would settle down into it, the side decks often disappearing below it, as if into a snowdrift. As the top of the wave lifted and balanced us, the full strength of the wind from which we had been sheltered in the trough, would strike us again, giving us forward motion even on the upward slope of the disappearing wave. Then again we would be in the trough and another great wave would sweep up from astern, hold over us the threat of its mass, and then lift us and swing us and drop us again. Then the next, and the next, and the next. They seemed to go on forever, and I wondered how it would end.

It always seemed to me impossible that the stern could lift to the racing green walls which were surely too steep to do anything except roll us

over or bury us under a cliff of water. The stern always did lift. Except
once. I was steering and had begun to feel confident in the boat and to be
sure that each wave would swing by below us. Then over my shoulder
I saw a wave that seemed to be higher than the others. Its slope was
steeper, and we were more underneath it. Its crest looked different, and
I saw that it was high and curved, and seemed to stand still in the air.
I had a flashing thought of a striking cobra. It hung and hung without
breaking. I saw then that it was going to break on top of us, and remember
thinking, 'This is where we go.' I wish that I had filled the cockpit with
sails. Now the cabin doors would splinter and the hull would fill.

Then the wave broke.

If it had waited and fallen six feet further forward, it would have filled
the cockpit and maybe the saloon. In fact only part of it fell on the short
stern deck and stern cabin top. The stern felt as if it were squatting for a
moment, and the green water piled up on it, frothing forward and along
the side decks. Some spilled into the cockpit, but not enough to fill more
than the bottom, and it did not even pull me away from the wheel. The
boat began to swing to port, but it was not hard to bring her straight
again, and the danger was over. The only damage was the tearing away
of the canvas dodgers, lashed along the rails and stanchions from shroud
to quarter. We were able to mend these later in kinder weather.

In the late afternoon the wind began to veer a little to the north and
the barometer fell slightly. I thought that we were in for another night
of gale and it might be worsening. We were all right to carry on. No one
had had much sleep, but we had all had plenty of rest, or as much as the
motion of the boat would allow. Peter was our best bad weather cook
and he had produced large mugs full of hot and solid stew. Even so, the
idea of this concentrated steering all through the night was not a cheer-
ing one.

Then in the evening the wind suddenly veered full to the north and
fell away to a light breeze. The sky cleared and the long seas started to die
away. By dark there was only a slight swell and we were ghosting along
in a wind that was 'very few'. Life came back to normal, and the last

twenty hours seemed like a bad dream, leaving no signs except the smoky saloon and torn dodgers.

About midnight the barometer began to rise from its low 1014 millibars. It rose to 1030 millibars the next day, and reached 1040 the following morning, a rise of three-quarters of an inch from the low reading of the gale.

The west winds we had been hunting had come in like a lion. They left like a lamb and that was the last we saw of them. We had a day of very light breezes from the north, and then three days of winds which were strong enough to give us a fair sailing speed, all from ahead, from the east and south-east. We put away the twin staysails and sailed very comfortably close-hauled, holding our course with the wheel lashed.

It was a joy to have finished with the rolling and to have a steady lee side, but we were going more to the north than we wanted. We reached the 40° north latitude and from the chart we saw that we now had two other problems to think about—steamers and icebergs. The nearest land was Newfoundland, less than five hundred miles to the north-west. I marked up each day's noon position on the American Pilot Chart which covered the North Atlantic and the Mediterranean. It gave no soundings, and was not used for navigation, but it brought the ocean to life, showing the flow of the currents and the directions and strengths of the winds which could be expected.

It showed the steamer lanes, and we saw to our surprise and dismay that the best course we could make was taking us straight for what appeared to be a meeting point for all the steamers moving between Northern Europe and all United States ports. There were so many converging black lines that I felt sure the sea would look like a busy harbour. Foremost in my mind was the danger of being run down, and I also wondered what effect the wake of a steamer might have on waves such as we had seen a few days before. Maybe a 'freak sea' could be formed, much bigger than the converging swells had caused.

In fact, we saw very few ships, either in this 'congested' area or anywhere else.

On 25th May, the day after our gale, we saw a large tanker going north-east. It was a little south of the steamer lanes and was some miles from us. We saw the next ship three days later, near the meeting point of the lanes. This one was much closer, and it taught us a lesson. From *Leiona*'s steering position there is a wide angle forward and to starboard which is completely blinded by the dinghy lashed on the cabin top. The canvas dodgers blanked out any view to the beam, at least to windward. We had a rule that the helmsman was to stand up every fifteen minutes or less and have a good slow look around, both to see that everything was all right on board and if there were any ships about. We usually sailed without lights, to save the batteries, but even when the lights were on they were low and small. I thought that our safety was better considered our own responsibility and should not be entrusted to the lookout or the radar of a fast-moving steamer. This ship confirmed my thought.

I woke suddenly a little before 3 a.m. No one had called me, and we were sailing quietly along. Michel was on watch. I had a strong feeling that something was wrong and got out of my bunk. I stood up in the cockpit, and saw over the starboard bow the lights of a steamer coming towards us, less than a quarter of a mile away. It looked as if she would pass on our starboard side, but only just. I said to Michel, 'How long has that steamer been in sight?'

He looked at me in surprise and asked in a bewildered way, 'What steamer?'

It was my turn to be bewildered, and I answered with some force, 'Good God, this one! Stand up and look! And steer to port!'

His mouth opened with amazement.

I was both frightened and furious, and said so, telling him that he could not have been doing his job and had obviously not been standing up for a look. He said that he had looked, and very often. I could only say that it was impossible for a blaze of light like that, red, green, masthead, cabin and deck to appear on top of us out of nowhere, and told him for God's sake to keep his eyes open and stand up more often.

Michel was not happy with my rebuke, and he must have spent a long time thinking about it. Next day he produced figures to prove that a ship coming fast to meet us could cover six or seven miles in fifteen minutes, and at that distance only the masthead light would be seen from our cockpit, and maybe not even that. His calculations were convincing, and when we timed ships later near the Azores, we found that they could come from first sighting to abeam in under twenty minutes, and these were not the fastest ships that could be met.

Michel had his revenge the next night. Our noon position put us just inside the broken red line on the Pilot Chart marked 'Mean Maximum Iceberg Limit'. I thought that like most government publications the chart was playing safe, and that the chances of meeting an iceberg at the edge of the limit were very slight. Even so, we were moving further into the area and the danger was worth thinking about. When I handed over to Michel at midnight I was still feeling angry over what I thought was his failure to keep a good lookout, and said to him, very wrongly:

'We are still in the steamer lanes so keep your eyes open. And look out for icebergs, too. We could meet one here.'

It was a stupid order, rather like saying, 'Battles will be won,' and Michel was too clever to accept it as it stood.

He asked what he would be likely to see if we did come close to an iceberg, all in a dark night. I could only answer, 'I really don't know.'

It had never occurred to me that we would ever be anywhere near an iceberg area, and I had not bothered to find out what could happen there.

'I can only think that if a mountain of ice arrived in this warm water, the air would start getting colder and that would bring along fog as well.'

This seemed a reasonable idea to both of us, so I went off to sleep.

A short time later I was awakened. It was Michel. He said, 'I think you had better come up and see what is happening.'

I got a shock. What had been a warm clear night had become damp and very cold, and in the starlight we could see that we were sailing into a bank of fog. A good breeze had turned into one that only just kept us moving. We slid into the fog and then we could not see more than two hundred yards, and probably much less. I thought to myself:

'This is really awful. The next thing that will happen is that we will bump into an iceberg, or a steamer will bump into us, which will be worse.'

I remembered Peter's tale of how *Didikai* had been run down in a fog off Spain. At the time of telling, it had been of interest, but remotely so. Now it was frightening to think about. All we could do was wait it out.

The fog cleared after two hours, and then came back again, or we found some new fog, shortly before dawn. Soon after we slid into the mist, we heard a foghorn, loud, and seeming to be quite close to the south-east, off the starboard bow. It came closer and grew much louder, drawing abeam of us, but never coming in sight. Then it went on and faded away to the south-west on its way to America. Not long after that the fog cleared, never to be seen again, and the sun climbed up bright and warm. A good breeze came from the south-west and we set a course straight for the Azores. We had finished with fog and cold and 'North Atlantic weather' and set off in the right direction with warm sea and warm wind.

CHAPTER SIX

Whales and Flowers

When we left the fog several good things happened at almost the same time. We completed our first thousand miles from Bermuda, we changed over to the chart of the East Atlantic, and we started to sail on a direct course to the Azores. And a steamer came and spoke to us.

The visit was a very pleasant surprise. During my watch on the morning after the fog I saw a steamer going the same way that we were, but three or four miles further north. It was about 8 o'clock. A little later I had another look, and saw that she had turned and was coming straight for us. By this time she was very close and I could see people hanging over the rails to have a look at this small boat miles from anywhere.

The steamer passed close under our stern and then straightened out alongside us, twenty yards away to windward. I had given the others a call and started the engine before our sailing wind was cut off by the tall hull and also to get moving at a speed which would give the bigger vessel steerage way. Thirty or forty people, passengers and crew, were looking down on us, just in time to see the sleepy heads of the young emerge one by one, two tousled men and two pretty girls. There was a lot of waving and calling out, but it was hard to hear anything over the noise of the engines and the steamer's wash. She seemed to be a cargo-passenger ship of about ten thousand tons, the *Concordia Cabo* of Haugesund.

The captain, in a very smart uniform of light brown gaberdine, came to the rail with a megaphone. I got ours out, and we had a shouted conversation:

'Do you want any water?'

'No, thank you very much. We are all right.'

It was a temptation to fill up, but we were not really short on our use of half a gallon each a day, and the process of getting water on board would be complicated.

'Is there anything you do want?'

'Yes, please. Could you give us a check on our position?'

I waved my sextant to show him what I meant.

'I will get that. Are you sure that there isn't something you want?'

We were completely unprepared for this sort of thing, and were not thinking either clearly or quickly. No one had any ideas. It seemed most ungracious not to accept this kind offer and they were obviously keen to help us. Then I thought of something.

'Yes, there is one thing we do want. We have a New Zealander on board and we have no beer for him. If you could spare us a few bottles it would be a great kindness.'

The captain spoke to someone. A minute later two sailors appeared at the rail of the well deck, below the bridge, with a line and some boxes. They slung two crates of beer down and we closed with them. I felt we were too close for comfort, but neither of us was rolling. I reached out with a boat-hook, hauled the crates onto the foredeck, slipped off the line, and we sheered away, calling and waving our thanks. The captain asked if we wanted to be reported to Lloyds, but I said that no one was expecting us anywhere so he wished us a good voyage and went off on his way to the Mediterranean. The position he gave us showed that my workings were right, which was good to know.

After they had gone we thought of things that we should have asked for—fresh meat, fruit or vegetables. It would have been nice to have them, but we were not in real need. We made a plan of action if the same thing happened again, and got out a list of welcome luxuries. It was a small list, because we were not in favour of being beggars on the high seas and it gave us a feeling of rather smug pride that we had everything we needed. Part of our plan was to show that we too could go 'pleasure cruising', and we were going to confront a daytime visitor with our bikini girls, banjo, rum punches, deck-chairs, and maybe even a miniature deck tennis set on the stern cabin top. Unfortunately, no other ship came near us.

It was a satisfying thought that we had no shortages a thousand miles from land, with the last port ten days astern of us and the next one maybe the same time ahead. Thanks to the cooks and to Murlo's stores, we were living very well and we were happy. Food played a big part in our lives. Breakfast was hot and sustaining. There would be coffee and tea, or cocoa if anyone wanted it. There was a choice of cereals and there was always some sort of egg dish with sausages, beans or bacon. Fruit and tomato juices would be on the table, and for the jam, honey and marmalade we had either biscuits, Ryvita, toast or scones.

The girls were very good at making bread and the weather had to be fairly bad to close down the bakery. Marie-Claire was always having trouble with the swinging table because she would forget that it was no good kneading the dough on one of the side flaps instead of the centre panel, and her first vigorous push usually caused a landslide. From the fog area until the Azores, we had a very peaceful passage and on many days the only entries in the log other than routine ones were domestic notes such as, 'Fresh croissants for breakfast (Murlo)', 'Crumpets for tea (Peter)'. 'Bread by Peter', or 'Apple and pear flan by Marie-Claire'.

Lunch was always a cold buffet meal, laid out on the saloon table about 1 p.m., following our noon rum punches. It consisted of bread and biscuits, cheese, meat paste, cold tinned meat or fish, and a salad of onions, cabbage instead of lettuce, and whatever other vegetables were still in use. After a while an extra meal crept in, afternoon tea, with toast or crumpets which were Peter's love and speciality. The afternoon also became the 'social hour' with the banjo and songs, and it led into the cocktail hour before sunset. Supper was the cook's big effort with soup, a meat course of some kind, and a pudding, fruit salad or cheese savoury. It was in the suppers that the skill of each cook was proved. They were all good, and each had a recognizable style.

Murlo was the most skilled and ambitious, and produced the greatest variety. Marie-Claire was often conscience-stricken that she had missed so many of her cooking days and put great effort into her productions. On her days we would have delicately flavoured fricassées and ragouts and good pastry. Peter's dishes were good and solid, all with a strong, but very welcome resemblance to sheep-farmer's stew. Michel was full of ideas, and had original and interesting plans for the evening meal. He would discuss these with us all, and the outlook became very exciting, but in the event the same dish always appeared, one that was known as 'spam schnitzel'. It was good, but familiarity bred, if not contempt, at least a lack of enthusiasm in greeting it.

We had some domestic problems which were interesting, anyhow to us in our forcibly restricted world. One was the mouse. Maybe there were more, but only one was ever seen. He must have joined at English Harbour. The first signs of a stowaway appeared on the shelf where I kept my navigating instruments. One morning I found that the soft eraser had been nearly eaten away, with narrow little tooth-marks all over it and a dust of small pieces of rubber scattered all around. Then attacks were made on bars of chocolate, although the robber seemed to prefer the silver paper to its contents. He did his marauding in the saloon but apparently lived near Peter's bunk where he would wake Peter by scampering over his bare feet. The attacks on food spread to biscuits and each day more food was being taken. We wondered if there were a thriving community of mice and what their food value to us would be if they ate everything except what was safe in tins. In Bermuda we bought some mouse traps and laid them in places we thought attractive to mice. We neither caught nor saw signs of mice after that, so maybe the threat was taken as a hint. We hoped it was not a case of the mouse deserting a sinking ship.

We also had a rat. I think that this time there really was only one. In a night watch after Bermuda, I saw what I thought was a black lump of seaweed showing over the lee dodger. When I stood up to look, the sea was clear but the dark lump seemed to move towards me. Then I saw that a large rat was walking along the rail. At that moment it was having a difficult tight-rope walk along the short chain between the two entry stanchions beside the cockpit. After the firm footing of the galvanized rail the swing of the chain probably upset his balance and this gave me time to lean across the cockpit and push him into the sea. His appearance worried me, because I thought there might be more on board, and I had visions of sails and ropes being eaten away. In fact, nothing was damaged, and we heard no squeaks or scrabbles, which I am sure a colony of rats would have produced.

Few boats leave the Caribbean without some cockroaches on board, and we were no exception. There were not many, nothing like the crowds I had seen in some boats where a galley table, suddenly lit up, would be a moving blackness, but there were a few. We put out poison in the form of boric acid mixed with confectioners' sugar, and this did kill a good many. It took me two years to guarantee that the boat was roach-free.

A problem which nearly brought about a crisis was a shortage of matches. When Murlo and I had worked out the quantities of stores we had reckoned that a box of matches a day would be more than we could possibly use. I had not foreseen how many matches even two smokers would get through. It was not the number they used for their cigarettes, but the amount of boxes that were left in the cockpit during the night, or on deck in the afternoon, and were made wet and useless. We only just missed arriving in the Azores living on cold meals or being driven to light oily rags from the Stuart-Turner spark plug.

When we had covered our thousand miles from Bermuda, changed to the chart of the East Atlantic, and set a direct course for the Azores, we felt that we really were coming within reach of Europe. For nine days we sailed this course of 95° without any alteration, and at the end of it raised Fayal fine on the port bow, exactly as we had planned and hoped. It amazed me to see our daily noon plot move straight along the pencil line of the course, and it amazed me still more when we made our landfall so exactly. Five different people had been steering, each with individual errors, sometimes feeling sleepy and letting the boat wander.

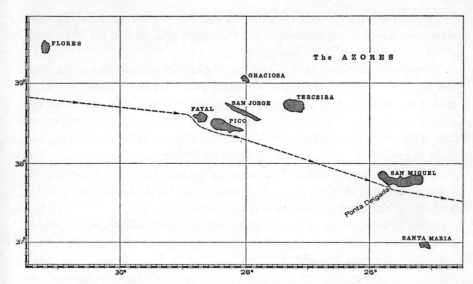

Fig 4 The Azores

As we neared the Azores, we began to notice more things. We first heard and then sighted one or two aircraft each day during the last week, and saw five ships in the last four days. The sea creatures became more interesting, and more alarming. We had the usual visits from porpoises, and would miss their quiet sighs if they did not appear alongside in the night watches. Nearly every day a pair of shearwaters would spend a few hours with us, and then disappear into nowhere.

When May turned into June, we had a repetition of the calm before Bermuda, and had to use the engine for two days. We motored close past a big red turtle which was too sleepy to do more than look dreamily at us. Murlo was the navigator by then. She loved swimming and had not been in the water since Bermuda. She asked if we could cut the engine for a while and have half an hour bathing after she had worked the noon position. It seemed a good idea until Peter suddenly stood up and said:

'Hey, look at that!'

Thirty yards away, moving with us, was the big black fin of the only shark we saw. He stayed with us for two days, and no one felt like swimming, even after the shark had gone.

The strangest things used to happen in my watches. On the night after the shark I was finding it hard to keep awake at the wheel. We were motoring, which made steering a course harder than with sails, and the engine noise was a soporific. The mainsail was up. I was startled out of my half-sleep when the whole sail suddenly shone white, as if a searchlight had been turned on to it. I looked around to see where the other ship was, but there was nothing in sight. Then astern, through the gap in the dodgers, I saw a streak of shining water. I first thought that it was our own wake and that I had been steering round in a circle.

When I had a proper look I could see that it was a wide swathe of brightness, while our only wake that showed at night was the narrow trail of the propeller, not two feet wide and petering out in thirty yards. This streak was about twenty feet wide and was shining with phosphorescence as brightly as a neon light. I thought it might be a column of jellyfish or some phosphorescent plankton, but then I saw that the front end of the streak was moving, and moving fast. That made it more likely to be porpoises or a school of fish. Then the head of the streak turned in a circle and came sweeping back towards us, rushing past the stern and off into the darkness to starboard. The front of the streak was squared and was followed by a swirling trail of light which reached up to the surface but did not disturb it. The edges of the trail were smooth and clear-cut, as if made by one large object and not by a lot of small ones. I could only think that it was a whale. There was no good reason to expect that it would harm us, but if it were only acting through curiosity, this rushing speed seemed odd.

I thought about calling the others, but decided not to. By the time they were on deck there would probably be nothing to see. If the whale did hit us, they would know about it soon enough. The shining ghost made two more passes under the stern, and then disappeared to the south. I had been startled out of my sleepiness and was keeping a better look-out.

About twenty minutes later the shiny streak appeared again, and this time it did run across our bow, very fast, passing right underneath. I slowed the engine down, left the wheel and stood at the port bow to see if I could make out anything directly below me. There was no more to be seen than before. This was its last appearance, and I was not sorry.

Two days later we were sailing along with a light south-west wind. In the morning I saw what looked like a tall white wave breaking two miles or so north of us. There were no waves big enough to make even white horses, but I thought there might be some more wind in the north. Half an hour later a white pyramid of foam rose out of the sea a mile to the south. Immediately, another column of white water burst upwards, and hanging in it could be seen a long black mass, at an angle of about 45°. It looked like a sperm whale, with a blunt nose and tubular body tapering away. We would have liked to be closer and have a good look, but not close enough to be frightened. The prospect of a hundred tons of whale shooting out of the quiet waters beside us was not a pleasant one, so we were glad to see no more.

At midnight on 5th June we were fifty miles south of Flores, still on our course of 95°. The wind had veered from south to west and then north as we came within the Azores. It was seldom stronger than Force 3, but we were doing more than a hundred miles a day with the help of a current which gave us a lift of ten or fifteen miles. We did not see Flores. All the islands of the Azores had high peaks, and I looked in the tables of Lecky's *Danger Angles* to find the distances at which we could first possibly see the 3,087 feet height of Flores, and the Caldeira of Fayal, 3,351 feet high, with the 7,613 feet of Pico fifteen miles beyond it. The geometrically possible distances were enormous. The two lower islands had a theoretical range of view of over fifty miles, and Pico could ideally be seen seventy miles away. We appreciated that conditions of visibility would always reduce these fantastic distances, but we were surprised how close we came before seeing land. The weather was fine but must have been slightly hazy and we did not see Fayal until within fifteen miles. That was in the afternoon, three hours after we had picked up the radio signal of Horta. We closed with Fayal in the evening, well before sunset, and had a beautiful sail along its southern slopes.

It was good to see land again and to know exactly where we were. It was also good to feel that we were in Europe, even if only on a remote fringe. Below the slopes of Caldeira a single road ran above the cliffs, and we could see the trees and small fields, the little villages strung along the road, and sometimes a car. Ahead of us was the huge volcano of Pico, but we only once saw its top above the white clouds piled around it.

Three fishing boats passed across us, gaily painted and well kept. The crews waved and shouted and pointed north, as if to show us where Horta was. I knew where it was but I had not planned to go there unless we had to, preferring to carry on to the most eastern island of the Azores, San Miguel, 150 miles away. We would then have a clear run for our last leg with no temptation to stop on the way. I had wondered if I would have any trouble with my crew over this. After eighteen days at sea they might find it tantalizing to be within a few miles of a good harbour and not go there.

I did have some trouble, but it was for the opposite reason. They did not want to stop in the Azores at all. Their argument was that the weather was good, and we were going along well, and we were not short of anything. Putting into port always seemed to involve more days than planned, and they were all for heading straight on for Gibraltar. Michel and Marie-Claire had no special reason to hurry, except to get Marie-Claire to the other side and relieve her parents of the anxiety she knew that they were feeling. I had arranged to meet some friends in Spain in the middle of July, but there was plenty of time for that. Murlo's mother had been pressing her to come to England quickly, so she had some sense of urgency. She was also thinking about her two days of sickness after leaving port. Peter was the only one who was really eager. A girl was waiting for him in England and at times he thought that maybe he had kept her waiting too long. All these reasons seemed good, but the argument for being able to start for the next thousand miles with full tanks of water and fuel and with fresh food was a stronger one, so we headed for Ponta Delgada in San Miguel.

The sun set as we left Fayal. The mountain disappeared in the darkness and all we could see were the four lighthouses, one after the other, as we sailed along the coast, and the twinkle of house windows, low down on our port side. At midnight we cleared the east end of the island and were again in the open sea. Under the high land the wind had been fitful, changing its direction and strength, but not quite enough to make us reef. In the open it steadied down again, but from the north-east, and we sailed close-hauled for San Miguel, at times being headed south of our course. Before sunset the next day we saw land, this time twenty-five miles away. The wind died away with the sun and we went on under engine. In the dark we motored below the high western end of the island, with the light of Ponta Delgada to draw us. We rounded the red light of the breakwater and moved into the harbour. The plan showed rocks and shallows outside the fairway, but there was one clear patch of two and three fathoms, just below what seemed to be the main square of the city. We anchored and shut off the engine. Everywhere was quiet, and nothing was moving in the harbour. As we waited for a while in silence, a clock in the square struck midnight.

We enjoyed our four days in Ponta Delgada. Early on the first morning the Harbour Master's launch brought one of the pilots out to us. He was full of apologies for not having sent someone in the night or earlier in the morning. Our midnight arrival had been noticed, but we had been mistaken for a fishing boat and reported as such. He showed us where we could lie further in the harbour, alongside a trading schooner at a dock. There were four pilots. They were most helpful, and after they had seen the two girls they could not do too much for us. They let us use the showers in the Pilots' House, invited us to drinks, asked us to meals or guided us to good restaurants and introduced us to the people of the island. Best of all, they drove us around and showed us the sights of San Miguel, which were very lovely.

We were told, as when visiting someone's garden, that we had come too early, and that July and August were the real months to see the island. Even so, it looked very green and beautiful to me, and the flowers were wonderful. I have never seen a place with so many, especially in the countryside. Some of the reason for this, I discovered, was that the road-men were not only responsible for keeping their stretches of road in repair but also for planting and tending flowers and bushes along the verges and grassy banks.

One day we were taken along the lovely cliffs and beaches of the north coast, which we were not going to see when we sailed away, and then over the ridge of the island to Furnas and its pretty little mountain lake, and back to the city, passing below the high peak of Barrosa. Another afternoon I was taken to a large plantation where pineapples were grown for export. The owner and his very charming family, and their assortment of dogs, showed me around the big formal garden, and after drinks in the palace-like house I was sent away with a load of pineapples for *Leiona*.

The most beautiful drive was the ten-mile climb to the Lagoa Azul—the Blue Lake. It is often a pity to make comparisons, but much of this drive brought back memories of Devonshire. At first, near Ponta Delgada, the trees and plants were almost tropical, with palms and hibiscus. Then we came into real Devon country; red earth and green fields, and neat little villages and white cottages, each with a tidy and colourful garden. The gardens overflowed into the lanes, and then the roadmenders took over so that we drove through an endless garden, flanked by roses, zinnias, stocks and lupins, and by hedges of fuchsia and great piles of blue and pink hydrangea. Even the people had a Devonshire look, thick-set and slow-moving with cheerful pink faces and rosy cheeks.

Sometimes we would meet a farmer's trap or cart, hard to pass in the narrow lane with its high banks. The fields as well had grassy banks instead of hedges, with rabbit holes in them. The grass was thick and deep green, dotted with buttercups, daisies and campions. In the fields boys were milking the cows, going to each one as she grazed, and not driving them into a shed. The cows were all standing in unnaturally straight lines, evenly spaced, apparently not tied up. This seemed a remarkable piece of training or behaviour until we saw that each was tethered by one leg to a stake and a chain, hidden in the grass.

9. Cala Figuera, Mallorca

10. Cala Pi, Mallorca

11. Castle and village, Cabrera, Mallorca
12. Puerto de Andraitx, Mallorca, in February

Our drive took us higher, and we climbed out of the green fields into highlands of short grass, heather and gorse, with brown stretches of bog. By now we were in cloud and mist, and we might have been on Dartmoor. We broke through the clouds and came to a bare mountain top of rounded spurs and ridges of grassy turf, broken by sheets of grey rock and small mountain pools and lakes, clear brown or bright green. As we came to a bend in the road our pilot stopped the car and insisted that we all closed our eyes until he told us to open them. We were led across the road and held standing for a moment. Then with great excitement we were told to look below us. It was a magnificent sight. We were on the precipitous edge of a huge, fertile, green crater—The Cauldron of the Seven Cities. Away to our left front, a thousand feet below us, was a large village with a church. It was the only 'city' unless we counted some other small hamlets or farms which lay in the bowl of the crater, around the shore of the big lake, two miles long and filling almost all the deep hollow. This time the resemblance was to the Nilgiri Hills of South India, even to the mists which were blowing towards us up the slopes of the valleys like monsoon clouds. The lake was strange in being of two colours. It was in two parts, a large circular expanse to the north, which was blue, and a smaller oval stretch to the south, almost separated by a long, narrow spit of land, which was a light green.

We had a language problem which we got round in a complicated but amusing way. None of us from *Leiona* spoke Portuguese with its slurred sibilants and broad diphthongs. Michel spoke some Spanish, strongly-flavoured with accents and phrases of the Yucatan backwoodsmen. Some of our acquaintances spoke a little French, and occasionally we met someone who spoke English. Our most used mediums were Michel's Spanish and, oddly enough, my French. Our most constant companion was one of the French speakers, and he found my imperfect and slower output easier to follow than Michel's natural and more rapid treatment. Michel would become very angry at hearing our guide say, '*Expliquez à vos amis, s'il vous plaît, que* . . .' It was surprising, but we seemed to get along all right, and we certainly all had fun.

4

The pilots had told us that we were the first yacht to arrive in Ponta
Delgada for two years, and that the last one had been a large motor yacht,
which hardly counted. Our fame lasted for only twelve hours. At noon
on the first day we were all together, walking along the promenade below
which we had anchored, when we saw a mast above the breakwater. It
looked too tall and fine to be a fishing boat. Then we saw the top of a
mizzen. No sails were set and she was motoring fast. Peter said, 'It must
be a yacht. It's a ketch.'

To Marie-Claire this was either magic or madness. 'Do not be so silly.
All you can see is two sticks on top of a big thick wall. How can you
tell what is behind that?'

We began to pull her leg:

'Yes, of course she's a yacht. A white one.'

'She has no bowsprit, but there is a very long counter.'

'Yes, but sawn off flat.'

'And a big doghouse amidships.'

'She had come from South Africa.'

And so on. By then she had rounded the breakwater. Marie-Claire was
astonished to see that our joking guesses were in fact correct. They were
obvious ones to make, except the last one. We were all astonished to find
that this was correct as well. She was *Stormvogel* on her maiden voyage
from Cape Town.

Stormvogel was an amazing craft, and it was an education to see her shape
and gear. There were fifteen on board, from Kees Bruynzeel, the owner,
to the four-month-old daughter of Gordon Webb, who had laid up his
Jenny Wren and was skippering and managing the boat with his wife as
cook and housekeeper. The crew were all young South Africans who were
paying to work and to have the tremendous voyage around the continent
of Africa, with the English ocean-racing season as an added attraction.
Some of them found our girls a counter-attraction. Murlo acquired a
shadow and I thought that we might sail with a bigger stowaway than the
rat and mouse which had left us.

Marie-Claire also had a shadow. Two German warships were in Ponta

Delgada, cadet training ships which had once been British frigates. We were invited on board and shown around. It was amusing to see that all the signs on the bell-boards and over the cabin doors were still in their original English. One of the young German officers was entranced by Marie-Claire. He attended to her every wish and took her for a second and very personally conducted tour of the ship. It was easy to see that many of the places he showed to Marie-Claire were not places of great interest, but in getting to them the maximum amount of handing down or pushing up was needed.

We were very sorry to leave the Azores, which had been very pleasant, and we sailed with happy memories. There had been only one snag—thick oil over the harbour water—but we were told that this was an unusual mishap due to a broken pipeline. The kindness of the pilots had allowed us to see places and people beyond our expectations and *Stormvogel*'s arrival had added to the fun. We were ready to sail on Saturday, but Kees and I decided to give our crews a dinner and dance ashore, and we had a tremendous time at a restaurant which was the fashionable Saturday night rendezvous. Maybe it was just because we were all feeling well, happy and confident, but that night everyone at the tables or on the dance floor was either beautiful or handsome. The last stars were fading as we walked back to our two boats in the dawn.

We had planned to sail before noon, and the pilots had said that they wished to come and have a drink of friendship with us, and to bring their 'Golden Book' for us to sign. They explained that this was a special honour, as the book was normally signed only by those above Commodore or equivalent rank. They were coming at 10 o'clock, and it was obviously going to be a fairly official visitation. I had a job to get the young out of their bunks, where they were making up for many hours of lost sleep, but managed to produce them, washed and dressed, but not breakfasted, in time for the pilots. The pilots arrived dressed in their best clothes and most pointed shoes. Politenesses, signatures and visiting cards were exchanged, and drinks were poured. Once the young had surfaced they functioned well. More drinks were handed out, the conversation flowed and the official farewell turned into a pleasant Sunday morning party. The pilots settled down happily and we all enjoyed ourselves.

By noon there was no sign that the party was ever going to end. I was becoming impatient to go, but did not wish to appear impolite to these kind people. I suggested that it would be a great favour to us if they would see us safely to the harbour limits, sailing with us and having their launch take them off outside the harbour. They thought this a splendid idea, and the launch was told what to do.

A west wind was blowing down the harbour, ready to take us out and on our way. While the young had been waking, I had got everything ready for sailing. Michel took in the bow line and gave a shove with his foot. Peter hauled up the jib, which pulled us round. I let *Leiona* swing on the doubled stern warp and then slipped it. No one had spoken, and we were away. Marie-Claire looked up.

'Keith, are we starting? Is this all? We are going now to cross an ocean! There must be something more to do before so big a journey!'

The pilots' launch did not begin to follow until we were clearing the breakwater. By then we had all sail set, and were moving fast. The pilots were delighted. After a while it was clear that we were going faster than the launch. It was a pity to waste a good sailing breeze, but we had to turn back and then heave-to until the launch caught us. We made our last farewells, and started to sail on with eased sheets. The launch steered alongside, holding off and not making fast, as there was a slight sea running. We had opened the chain 'gate' in the rail and the pilots stepped out backwards, one after the other, into their launch.

The last one was just letting go when he remembered that he had not kissed the hands of the ladies. He hauled himself back on board, holding the briefcase of the Golden Book to complicate things, and implanted his kisses. Then with a ringing farewell, he stepped boldly out for the launch. Peter and I grabbed him just in time, because his coxwain had had to sheer away, and he was about to put himself and his book confidently into the sea.

We had a good sail along the south coast of the island, which I hope to see again some time. At sunset we had the eastern end abeam of us, and by midnight the bright flashes of the Ponta de Arnel light had dropped away.

Cape St Vincent was our next target, eight hundred miles away. We wondered how many days our passage would take and what adventures it would bring.

CHAPTER SEVEN

Into the Mediterranean

West winds took us away from the Azores and by sunset the horizon astern was empty. We were on the last leg of our journey, and the next stop would be the final one, the end of the voyage and the breaking up of the crew.

I had always felt that we would find our strongest winds and probably our greatest troubles in this last stretch, but I was wrong. We had a wonderful sail to our landfall. The winds behaved according to the weather chart. For the first two days we had light winds from the west, and made fair progress with the mainsail guyed forward and the genoa jib boomed out with one of the running-sail poles. Two days of freshening winds followed, varying between north and west. Then we came into the Portuguese Trades and really began to sail. The wind was from north and north-east, steady and strong, pushing big long seas before it. We had to put six rolls in the mainsail, keeping on the staysail, which is a small one, and big No. 1 jib. This was probably a bit too much sail for *Leiona*, who is comfortable and steady rather than fast, but we had two thrilling days, with spray flying over us and green water on the lee deck. We sailed 150 miles one day and 151 the next.

From the Azores we had set a course slightly north of our direct route, heading towards Lisbon, so that we would not have to work up to windward against the Portuguese Trades when we found them. This was a help now, as we were able to ease away to the south with these strong winds pushing us on from just abaft the beam. The big jib was the one really doing the work, pulling and lifting us forward and upward so that we seemed to be carried over the waves as they swept under us.

The biggest thrill was steering at night, alone in the cockpit, hooked on with a safety harness, the rest of the boat in darkness and sleep. The black sea to windward would swell up against the dark grey sky, growing higher as it rushed nearer, showing a line of frothing white as it raised itself above the boat. Then *Leiona* would heel over and be lifted up, seeming to gather speed as the crest of the wave hit her side with a thud and a splash of spray, and then a mass of seething white water would hiss away into the darkness to leeward.

Cape St Vincent was our landfall, and we first saw it in the early afternoon of the 18th of June, five miles away to the north-east. I had a feeling of relief but not, as I had expected, of triumph. Instead, there was a sensation of anti-climax and almost of regret that the easy and pleasant routine of the past few weeks was to be exchanged for the complications of an arrival on land. The others must have felt rather the same. When I said, 'Well, we should be in Gibraltar the day after tomorrow,' there was a silence. Then either Murlo or Peter said, 'What a pity it is ending. I wish that it could go on for longer.'

And Michel turned to Peter, saying, 'Come on, Peter, let's have a quarrel. No one has had one yet, and I thought that they always happened in small boat voyages.'

In fact, neither our journey nor our troubles had quite ended. Just before we sighted St Vincent, our good north wind began to die away, and by the time we were under the lee of the land there was a flat calm. We turned on the engine and thought that we were in for a night of motoring. As the sun set a wind got up from the east and began to increase in strength. During the night it veered to the south-east and began to blow hard, with a very short, steep sea. We could no longer hold our course for the Strait of Gibraltar, and had the choice of heading either south or east. We went east, almost on a course for Cadiz, hoping to get some shelter as we neared the land, if the blow went on as long as that.

Close-hauled, under only reefed main and staysail, we were making very little progress against the steep seas, and had to help the boat along with the engine. It blew all night and all the next day. Poor Marie-Claire had taken to her bunk in the fore cabin, and that was where the noise and motion were greatest. I could tell from Michel's questions that she was becoming frightened as well as sick, and was longing to get to land. Michel asked if I had decided to go into Cadiz. I told him that at 9 p.m. I would know what to do. By then we should be a few miles outside Cadiz. If things were no better we would go in, but if the wind had eased and the land gave us shelter, I wanted to turn south and carry on. Michel, probably speaking for Marie-Claire, urged me to go in anyhow, but I would not change my plan. If we went there we would be sure to stay for at least a day and maybe several, and by now I was eager to make Gibraltar our next stop and not to waste time. Also, from the chart, the entrance did not look easy and I was not happy about trying it.

At 9 p.m. the lights of the harbour were in front of us, a few miles away. The wind and waves were still giving us an uncomfortable trip, but they were getting less. I decided to go south. We tacked and put up the jib, and cut off the engine. We went along well, doing five knots, but with a lot of spray, crashing into the waves and not swinging over them as we had off the Portuguese coast. At midnight the wind got stronger and the movement became violent. Michel asked me to turn back. I felt very sorry for Marie-Claire, but I went on, taking off the jib. Things were easier then, but slower. Still, we were not racing, and safety and comfort came first.

As morning came the wind dropped and became 'very few', leaving a big swell which also died away as the day went on. It was a temptation to finish our journey as we had started it, under sail and no engine, but it would have meant a slow beat to windward and another night at sea. We swallowed our pride and motored on. Before noon we closed with the land at Trafalgar.

Our evening rum punches were drunk as we passed Tarifa Point, and there we celebrated our arrival in the Mediterranean. We entered Gibraltar Bay just before dark, over a glassy sea. It was my first visit there, although

I had seen it many times. I had more of a feeling of excitement here than at St Vincent. This really was the end of the journey.

We turned in through the breakwater and hailed the Harbour Master's office. They told us that it was too late to dock and that we would have to wait outside until the morning.

Murlo called out very plaintively:

'Can't we really come in? We have come an awfully long way.'

A few more questions and then we were guided by a launch to a berth at the north end of the harbour. As we drifted alongside, a small round figure standing on the dock above us called out in a cheerful voice:

'Is that *Leiona*? Colonel Robinson? I have some fresh bread and milk, and some mail for all of you. I am Charlie.'

This was my first introduction to Charlie Rodriguez, the most happy and helpful of all yacht agents. I had been told of him in English Harbour, and advised to write to him. Gibraltar seemed a long way from Antigua, and I thought it was too much like tempting providence to give a written forecast of *Leiona*'s expected arrival. But the Azores seemed within reasonable distance, so I wrote from Ponta Delgada. By then I had forgotten Charlie's full name, so I addressed the letter to 'Charlie ???, Yacht Agent, Gibraltar', and hoped that it would find him. I think that just 'Charlie, The Mediterranean' would have been enough.

Our days in Gibraltar were busy, interesting, amusing and yet in a way sad. Until then, our interests had been common interests, in the boat and in each other, and they were concentrated rather than distracted. Now our private lives took first place. Everyone had letters to read and write, and future plans and moves to think about. Marie-Claire's parents, a count and countess straight out of a romantic novel, drove down from France to greet her and take her and Michel home. They stayed long enough for us to get to know them with affection, and for their younger daughter and Peter to become admiring friends even without a common language. I had some army friends there, and through them met some of the people, English and Spanish, living in and near Gibraltar. Everyone we met was kind to us, and we were busy with visits to their houses for meals or baths, or with their company on board *Leiona*. There were afternoons at polo matches, and picnics, and a bull fight.

Michel and Marie-Claire were the first to go, and then Peter, who flew to England. Murlo was going to stay for the first part of my move up the Spanish coast, but I was held up waiting for new headsails from England, and her mother was pressing her to return. A very fast motor yacht was about to leave for the French Riviera and they were glad to take her that far, which would, at least, reduce the cost of travel. So she went off, too.

Once again I was alone in my boat. I like having my friends to sail with me, but there is always a feeling of relief when I find myself on my own. I feel guilty about this ungrateful attitude, and, in fact, I am always delighted to welcome my visitors and glad to have them with me, but it is still quite a joy to be alone in a boat. Everything is quiet and there is plenty of room in which to move around, to lie about, or to put things. And it is easy to get on with the never-ending small jobs a boat demands without being deterred by the fear of disturbing sleepers or sunbathers, or being distracted by some more amusing alternative.

Being alone, I had more time to think about the past and the future. This journey across the Atlantic had really been only a preliminary to my new life. It was going to take a few more months, or maybe some years, to see if I really did like living in a boat, and if the pleasure and freedom she gave outweighed the restrictions and discomforts, when compared with normal life ashore. The biggest thing that the Atlantic crossing had taught me was that I knew very little about boats and that I had more to learn than I had thought possible. I had been lucky in having Peter and Murlo with me, profiting from their knowledge and experience, and I wondered if I would be able to cope with the problem of looking after a boat, doing most of the work myself to avoid the expenses of shipyards and paid hands. I wondered also if I would be able to keep the boat and myself on the small income of an army pension. The next two years would show me.

Anyhow, there were more things to do than just to sit and think rather gloomily about possible troubles in an unknown future. Some new

Terylene headsails I had ordered long ago had not yet arrived from England. I discovered after some searching that they were ready and waiting but had not been posted. So as not to waste time while waiting for them I moved round to the Yacht Marina and with two excellent Spanish fishermen, brothers from Algeciras, I started intensive work on varnishing the mast and rails, and repainting the topsides and deck. This work was held up by bad weather, which I had not thought possible in the middle of a Mediterranean summer, caused by several days of the Levante, an east wind which funnels down the Strait bringing rain and fog and a generally depressing dampness. It would announce itself in the early morning by a long plume of white cloud curling over the top of the Peak from the east and trailing down the western slope. It would put a stop to outside painting, and we would try to find some other jobs to do.

These delays were irritating because I had asked some English friends to meet me in Alicante on the east coast of Spain, in July, and to come sailing from there. This arrangement had been made in England in January, and it seemed an easy date to make at the time. I had hoped to reach Gibraltar in mid-June, as in fact we did, and the 250 miles more to Alicante appeared so small a distance compared to the thousands of the Atlantic crossing that it seemed no problem. In Gibraltar I began to think differently about it. I had hoped to leave Gibraltar before the end of June, but sails, painting and Levantes were obviously going to delay me. Also, I found that I was feeling much more worried about the short passage to Alicante than I had been about sailing over an ocean.

I was beginning to learn something about sailing now, and so was losing the bold confidence of ignorance. My courage was further shaken by hearing about the unpredictable Mediterranean weather, the unheralded squalls of great violence, and the bad conditions likely to be found off Cabo de Gata, 150 miles east of Gibraltar. And I did not like the idea of being so close to land, and in a focal point for all the shipping going into or out of the Mediterranean, especially as I was going to be alone.

Luckily, as so often happens, everything turned out well. Richard

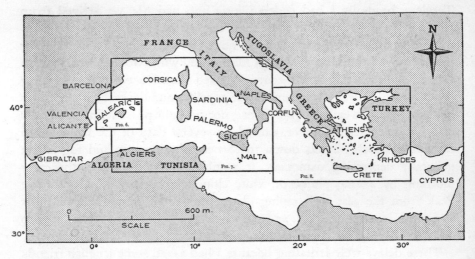

Fig 5 The Mediterranean

Gardner cabled that he and Diana were arriving in Alicante on the 16th
of July and not the original date of the 12th. This gave me a few more
days to make the rendezvous. Some good, dry, westerly weather appeared
so that we could finish the painting of the deck, and my new sails came
by air. One afternoon a sailing dinghy stopped alongside and a young
Englishman called out that he had heard of my trip up the Spanish coast
and would very much like to come as far as Alicante. We started off on
the morning of the 8th.

Nicholas Prosser was a good sailor and a good companion. He found
his way easily around the boat and we got on well together. We had a
very peaceful sail to Malaga, sixty miles along the coast, where we arrived
late at night. We hove-to outside the harbour and went in after breakfast.
The most amusing thing about the day's sailing had been the numbers of
dolphins, porpoises and big blackfish which were with us until sunset. I
have never seen so many. They seemed to enjoy playing around the boat
and we could see great schools of them coming towards us. For the first

time I saw really young ones, keeping close to their mothers and often sandwiched safely between two large ones, almost being carried along.

Malaga was my first Spanish port, and I enjoyed it. We left *Leiona* in a quiet corner of the harbour nearest to the city and wandered ashore. There was mail to collect from the bank, and more from the farm of a friend a few miles away in a mountain village. It was nearly dark when all this was over, so we decided to wait until morning to sail. We dined in a restaurant on a high peak above the city and watched the night creep over the harbour while the lights of the roads and houses came to life.

Except for the last few miles, the rest of our sail to Alicante was pleasant and without trouble. I was surprised to find how little we saw of the coast and the high Sierra Nevada mountains, close to the sea. In this fine midsummer weather the land was masked from the sea by a dusty haze like a brown mist. Only the mountain tops could be seen and not all of them. I had thought that by day it would be easy to fix our positions by bearings from the mountains, but I found that it was very hard to identify them, and to tell if an isolated peak standing out of the haze was a high one far away or a low one nearby.

Cabo de Gata produced no bad weather, and we had to motor past in a flat calm for a couple of hours. I had wanted to see what Cartagena was like, but I felt that time was short, so we sailed past it in the early morning, hoping to get into Alicante late that same night.

Instead, we spent the night at sea. As we sighted the hills above Alicante, in the late afternoon, a strong north-west wind started to blow. A steep sea began to get up quickly, and before sunset a gale was blowing.

By then we were making slow and very wet progress northward, under staysail and deeply reefed main. I wanted to see what it would be like hove-to, so we hauled the staysail to windward and lashed the wheel. *Leiona* lay very comfortably, facing north and being slowly pushed to the north-east, safely away from the land.

It was my first time of heaving-to in bad weather, and I found it interesting and encouraging. It was quite different from the nights across the Portuguese Trades. The same white-crested black waves were sweeping down from windward, though not as big, but there was no spray and no thudding on the side. We stayed in the same place, moving easily up and down as the white water seemed to pass beneath us and lose itself in the darkness. It was an impressive night. A hot wind was screaming in the rigging and the waves were roaring and hissing as the crests were tumbled over by the gale. Venus and Jupiter were bright in the moonless sky, one to the east, one to the west, and beyond our port bow a whole mountain was on fire. In the dusk a few red and orange clusters of flame had appeared on the lower slopes of a high mountain north of Alicante. The fire grew with the night as the gale fanned it. All through the night it stood against the grey sky, a glowing pyramid of red and yellow, the flames still showing as daylight came, when the red glow changed into a streaming mane of grey smoke.

The gale went on blowing the next day until two hours before noon, when it stopped as suddenly as it had started. A light breeze followed it and took us gently into Alicante five hours later.

We sailed into the big clean harbour and anchored in front of the Club Nautico. Nicholas went back to Gibraltar by road the next day, and I stayed to wait for Richard and Diana.

This seemed the real end of my Atlantic crossing, when I was properly in the Mediterranean, with my boat all ready for the appointment which I had made with Richard so many months before. Now I was really about to start my Mediterranean sailing and begin my new life, one of living entirely in a boat. So far, I was very happy, both with my life and my boat.

CHAPTER EIGHT

To the Balearics

Alicante was a pleasant place for waiting, and I was happy to wander about and make my first acquaintance with Spain. I liked the wide streets and the tall buildings with their changing contrasts of bright sunlight or dark shadow. My favourite place was the wide promenade along the front, set between high palm trees. Its twisting patterns of coloured tiles were watered each evening to be cool and shining for the *paseo* of the town's young people.

My days started with the Spanish business man's breakfast of coffee, a croissant-like *ensaimada*, and a shoe-shine, all at the same time at a table beside the sunny promenade. The people in the streets and the shops were kind and helpful to me in my slow efforts in their language, and never laughed at my faulty grammar and strange pronunciation. To my surprise, an involved explanation and a simple drawing in the dust of the Club Nautico's flat roof resulted in a well-fitting sail for the dinghy, made by the club's boatman.

I enjoyed my stay, but I was looking forward to my crew's arrival. Richard and Diana were driving to meet me. Richard had been a soldier, and now he lived in a cottage in a quiet Dorset village, keeping himself amused with fishing and golf. I was glad that he did nothing else, because he was one of the few of my friends with enough spare time to come sailing with me.

Diana was a very welcome addition to the boat. She brought decoration and entertainment, and the feminine touch which made the boat happier

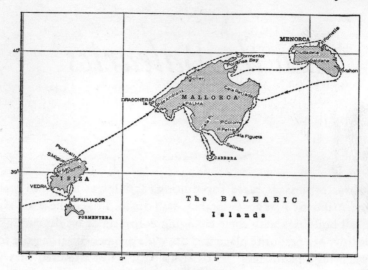

Fig 6 The Balearics

and more comfortable. She took over the galley and fed us well, but she was not what everyone expects a ship's cook to look like. She was tall and fair and pretty, and her real occupation was painting, which was her profession. A lot of the fun of our sailing came from finding places for Diana to paint, and anchoring to suit her idea of colour and light and composition.

A dusty car stopped on the water front and two waving figures called me and the dinghy to the shore. We unloaded the car, found a place to leave it in safe care, bought a stock of bread, eggs, fruit, and wine, and sailed in the afternoon, heading for the Balearics.

I had cleared my mind of one problem, that of bringing *Leiona* across the Atlantic, but now I had another one, of finding a place to spend the winter. I hoped that our next few weeks of sailing might solve the problem.

From conversations, books, and charts I had formed the idea that the Balearics would provide a good summer sailing ground and a winter home. I had learned that the Mediterranean in winter was not quite what the travel advertisements tried to make us believe. I could expect blue skies for most of the time, but there would probably be two or three really bad gales each winter.

All this applied to the Mediterranean as a whole, except that some areas in the north might have a greater share of cold and wind. The Balearics seemed to provide as good a climate as anywhere, and the group of islands with many inlets and channels was more attractive for our summer sailing than the long almost unbroken coastlines which bordered the western Mediterranean. While we were sailing among the islands I would have time to look for my wintering place.

I had a vague idea of what I wanted and I could see that it would not be easy to find. First of all I wanted safety from the winter gales. This could probably be found in most of the main harbours, but so would noise, smoke, bustle, and dirty water. I wanted to avoid this, and anyhow I liked village life better than city life, so a small village harbour was the second need. The third need was a place where I could get food and water easily, and where there were carpenters and mechanics who understood boats. And finally, I wanted the place to be beautiful.

So each time we sailed into a new bay or harbour in the Balearics we were looking at it with different thoughts. Diana was wondering if she would find a picture to paint. Richard was trying to guess where I would tell him to let go the anchor. And I was looking to see how the features around us fitted my needs for a winter home.

Ibiza and its islands did not produce the winter home, but we had two weeks of happy sailing there. This was the first time that either Richard or Diana had been in any sailing boat larger than a dinghy. My invitation for a summer of Mediterranean sailing, given in front of their Dorset fireplace on a cold and rainy winter afternoon, had been accepted with delight. As the summer came nearer they began to have misgivings over their lack of knowledge about boats. They bought books and asked friends and did everything they could, except going sailing. By the time they arrived in Alicante they were so full of confusing facts and theories that their misgivings were greater than ever, and they were almost sorry to see *Leiona* waiting and ready for sea.

Luckily, the weather was good, and *Leiona* was a good teacher. We sailed out of the harbour and found a quiet sea and a light south wind, perfect for taking us towards Ibiza, a hundred miles to the east. I could see my crew being very quiet and thoughtful, watching the land slowly fall away, and wondering what it would be like when we were alone in the sea, and in the dark.

Richard had steered us out of the harbour, and after a while he began to look more cheerful as he found he could control the boat without trouble. I started to make out a watch list, and asked Diana if she would mind steering for an hour in the middle of the night, to give a little extra sleep to Richard and myself.

She was horrified.

'No, Keith, I couldn't do that. Do you mean I would be all alone here, trying to steer the boat in the darkness? What an awful idea!'

I just managed not to laugh at her dismay.

'I promise you, Diana, you'll be all right. Try steering now to get the feel. Richard has been doing splendidly. Everything goes the same way, the wheel, the boat's head, and the black line on the compass.'

Her eyes opened even more widely.

'A compass? No, this is too much. I won't know where to go and we will get lost.'

'No, we won't. Anyhow, not for long. And I shall be sleeping only a few inches away if you need me. So have a try.'

She did, and she enjoyed it, sailing alone in the stillness of a clear night. After that neither she nor Richard were worried about what they didn't know, and they let *Leiona* teach them.

I was longing to get into the Balearics. Three months ago in the West Indies, which now seemed so far away in distance and time, my charts had been marked with the names of places and people. One of these places was just south of the island of Ibiza. It was called Espalmador, and it was a very small island, almost joining Formentera, the most southern of the whole Balearic group. The anchorage that had been marked showed on the chart as a circular lagoon a quarter of a mile wide, surrounded by a ring of low hillocks, a long sandbank, and a thin reef with a single entrance from the south-west. We planned to make it our first landfall and anchorage.

In fact, we saw something else first. We were looking out for Espalmador in the hot haze of the afternoon, when over the port bow appeared the unexpected outline of a skyscraper, quivering in the mirage of heat. Soon the mirage began to clear, and we saw that our skyscraper was not a building, but the tall rock of Vedra Island.

Vedra stood up suddenly out of deep water, a jagged spire of grey and brown rock more than a thousand feet high, rising vertically from a small base. We had seen it from its narrowest elevation, where it looked like a single pillar. As we left it five miles to the north we began to see the low curves of Formentera and Espalmador ahead.

An hour before sunset we found our way into the lagoon. It was perfection. Around us was a ring of white sand, uncovering a comb of black rocks to the south. In the east the sand formed a causeway narrowly separating the still waters of the lagoon from the waves of the open sea. The northern half of the ring was backed by a lumpy ridge of dunes and small hummocks covered by long grasses and a few bushes and low trees.

We anchored in three fathoms of clear water, over a bottom of white sand and patches of dark green weed.

The cool of the evening had taken away the earlier haze of the day. Above the grassy ridge Vedra's spire, ten miles away to the north-west, stood like a grey shadow, and a new sight appeared to the northward, nearly as far. This was the white citadel of Ciudad Ibiza, the city which gave the island its name. We were delighted with the quiet and beauty of our first anchorage.

We had the water to ourselves, but there were two houses on the ridge. One was a small grey cottage, near which was standing a man watching a flock of sheep. The other was a long, low, white house, hidden from the entrance by the curve of the ridge. We put the dinghy overboard and went ashore to explore the tiny island and to find out if we had trespassed or were unwelcome in a private bay.

As the dinghy touched the shore, fringed with a thin line of pale pink sand, the shepherd's dog came to greet us. She was a true Ibizencan hound, a breed dating from at least as far back as Carthaginian times, and whose pictures can be seen on Greek vases two thousand years old. She was a happy, gangling, half-grown puppy, the size and shape of a Scottish deer-hound, but with a thinner coat of rough white hair that was nearly pink, and with patches of red-brown. We had been talking about these dogs, and it was strange that we should meet one the moment we landed.

The shepherd seemed glad to see us, and we had a long and laboured conversation with him. At the end of it all, our impressions were that the bay was free for anchoring, the shepherd lived there all the year round, and the island belonged to his master, one Don Bernardo, who had come from Barcelona for his summer visit. By the time we had worked out all this information dark was falling, and we rowed back to *Leiona*.

Early morning in the lagoon was as beautiful as the evening had been, and we decided to stay there and paint and swim and walk. We were enjoying the warmth of the climbing sun when a small boat motored through the entrance and went to the wooden pier below the white house. A dignified gentleman with a white beard and a broad brimmed grey hat came from the house and got into the boat. The boat left the pier and made for the sea, sweeping near us on its way. I stood up and began to think out my most polite Spanish phrases. I had only got as far as saying '*Buenos dias!*' when I was answered in perfect and unaccented English:

'Good morning. I am very glad to see you here. I hope that you will all come to my house for coffee and brandy tonight at about ten o'clock. And my grandchildren would very much like to visit your yacht this afternoon.'

We exchanged the visits and enjoyed the good company. Don Bernardo was an Irishman who had married a lady of Barcelona and had made Spain his home. He had built this house on the deserted island and came for most of the summer months, leaving the shepherd to look after it in the winter. The house was filled with his family and their friends. Their own boat brought them food from Ibiza, or took them there for the day if they felt their paradise was too lonely. Each Thursday the boat brought a priest to stay the night and to hold confession, mass, and communion in a chapel built in the garden.

It was something of a shock to hear that they had so mundane a difficulty as servant trouble, but this was their only worry, that the servants from Barcelona did not like the food of Ibiza or the remoteness of this wonderful island.

Three days later we left Espalmador, sad to go, and sailed up to the main island. We went first to the big harbour of the city before starting off on ten days of great variety among the small villages and coves.

Ciudad Ibiza was an impressive port to enter. We passed close by the rocks of Malvin and Dado Grande to sail beneath the citadel and then to round the high breakwater and see the old city closely packed on the steep hillside to port, while the wide flat valley stretched away to starboard. It was a great change from Espalmador, and we were not as happy made fast to the windless breakwater as we had been in our clear and empty lagoon.

But all the same, it was fascinating to be in Ibiza, either among the houses crowding the narrow streets, or along the dock busy with trading schooners, fishing boats, steamers, nets, carts, trucks, and people.

The city was at its best at night, away from the sea front and in the dimly lit streets and alleys which wound to the top of the hill. The old white houses were strangely shaped inside and out as they were fitted somehow into the sharp corners and steep hillsides. Our midnight walks within the gates of the old city walls left us with thoughts of cloaks and daggers, masked ladies, troubadours, and link-boys.

We left the city and sailed up the west coast of the island, in perfect weather, each evening giving us a contrast of anchorage. We passed through the narrow channel between Vedra and the cliffs of the shore, close under the pinnacle that had been our first sight of the Balearics. We anchored in the wide harbour of San Antonio's tourist town, and off the wide sandy beaches outside it, and in the square cove of Cala Badella, so well hidden that we nearly missed it.

We sailed and lazed and swam our way along the short but beautiful north-west coast, where peaks a thousand feet high stood over the sea, and the shore line was broken by jutting headlands and steep valleys. We learned that a *cala* was a bay of any sort, maybe an open one, and that a *puerto* was not a 'port' as we think of it, with docks and stores and cranes, but just an enclosed bay, maybe deserted, but almost certainly a sheltered anchorage.

The vineyards of the deep *puerto* of San Miguel kept us for two days, and we left ashamed that we had been too lazy to walk up to the little church and village at the top of the valley, only a mile away.

The low grey cliffs and overhanging pine trees of Portinaitx encircled our last anchorage in Ibiza, where we shared the forked end of the cove with two Valencia trawlers, lying to stern anchors with their flaring bows made fast to the rocks below the white taverna.

On the last afternoon in July we set off on the fifty-mile crossing to Mallorca, again with a light south wind to take us to the north-east. We had our perfect weather again, and with a steady barometer we felt sure of a gentle sail, even if a slow one. Then the wind played its first trick on us.

In the afternoon I had called Richard from his siesta to see some dolphins astern of us, leaping high out of the water. I quoted the old sailors' saying:

> 'When the sea-hog jumps
> Look to your pumps.'

We both laughed, saying that it must be a North Sea saying, not applying to the Mediterranean. We laughed too soon.

An hour or so later I was still steering, very sleepily. There had been no change in the sea, nor in the light off-shore wind. In one of my looks around the horizon I was startled to see some black cloud out to sea, far away to westward, almost astern. It was well to leeward, and as I reckoned on clouds travelling with the wind and not against it, I was interested but not concerned.

A little later I had another look. This time I could see the cloud more clearly, and a lot more closely. It was long and thin, shaped like a drawn-out sausage, inky black, and stretching over a front of three or four miles, with what looked like dark grey sky beyond it. It looked a very peculiar cloud, and seemed to be moving towards us like a rolling pin, twisting over and over as it came. I suddenly thought, 'Line squall!' I had heard and read of such things but had never seen or really imagined one. I still did not understand how it was moving against the wind. Two years later a paragraph by Errol Bruce made it clear; the squall was not moving against the wind, but pulling all the surrounding air into its own turmoil.

I kept on watching the squall. There was no swell, and even with the binoculars I could not see any waves below the cloud. Then I suddenly saw that we were in trouble, and maybe in danger. The line of cloud was now very near us and moving fast, so fast that there had not been time for swell or waves to get up. Beneath the black line was a floor of white water, whipped up and laid flat by the force of the wind, which was almost on us. I shouted:

'Richard! Diana! Quick! On deck! Get the jib down, and hurry!'

I slacked the mainsheet right away and went forward to get the staysail down, hoping to get the main off as well before the wind hit us. I didn't know quite what to expect, but I thought that the less sail we had up for a start the better.

I had been too slow. Richard and I were finishing our hurried and untidy stowing of the sails, and Diana was on her way to the wheel, when the wind arrived like a wall and laid us over flat to starboard. I got back to help Diana at the wheel, and we managed to head *Leiona* down wind, which let her sail at a safer angle, but too fast.

Richard and I tightened in the mainsail a little, and rolled eight turns into it, which made it look the size of a storm trisail. Then I thought about what to do next. The course we were on was taking us just clear of the northern tip of Ibiza, but I was afraid of wind changes which might drive

us towards land. So I brought *Leiona* slowly round, until we were going
to windward.

The wind was still just as strong, and now there were some waves with
it, not very high, but short and steep. As we came beam-on to the seas a
lot of spray and some green water came on board, but as we came further
into the wind things got much better. I was afraid that with only the
mainsail set *Leiona* would 'weathercock' and point dead into the wind,
becoming unsteerable. To my relief she hove herself to beautifully, probably helped by the windage of her high bow and the untidy bundles of
the two headsails. We lay heeled gently to starboard, getting some spray
over us, but no solid waves, and we were pushed slowly northward, which
was a safe direction into the open sea.

This went on for three hours. The wind lessened, but the waves built
up steeply to a height of six or eight feet. Then about midnight the wind
died away and left us in a clear and windless night, but with a big swell
which threw us about more roughly than the wind-waves had done. We
started the engine and motored off to the east, twisting and pitching over
shining mounds of swell.

Mallorca greeted us at dawn with another tower of rock, nearly as tall
but not as jagged as Ibiza's Vedra. This was Dragonera Island, off the
western point of Mallorca. We saw first the flash of its lighthouse and then
its bold outline. It was just the shape of a dying dragon in a fairy story,
with a long humpy back sloping down to a thin neck and a small head
stretched out to the north-east.

Dragonera was our mark for finding Puerto de Andraitx, another small
bay which had been marked on my chart, with the words, 'Good harbour.
Unspoiled. See Bobby Somerset.' If it had not been for this note we might
never have gone there, as the harbour was hard to see on our chart and
was well hidden from the sea. We wondered how to pronounce the name,
with its odd Catalan ending of '-tx', and all our guesses were wrong. Later
I was given the simple guidance of calling it 'Andraitx like scratch', which
was exactly right.

Puerto de Andraitx was a pretty little village at the end of a long narrow

bay running straight inland between two high ridges. The bay was open
to the west, but two breakwaters had been built, one near the mouth of
the bay, and one near the village, forming an inner harbour. We sailed
right in to have a look at the village, and then went back to anchor in
the empty pool inside the arm of the outer breakwater.

Of all the places we had seen this was the first which fitted in with my
idea of a winter home. It had a small, very rough, slipway big enough to
take *Leiona*, and it looked as if I could anchor safely in either of the
harbours, or make fast at the inner breakwater. There was plenty of time
to look at other places before deciding, but already I had a liking for
this one.

It also seemed a guiding coincidence that Andraitx should be the winter
home of Bobby Somerset, and that his name should have been written on
my chart. I had often read and heard about him and of his voyages in
Jolie Brise and *Iolaire* and *Thanet*, and I very much wanted to meet him.
I also wanted to ask his advice about Puerto de Andraitx, now that I had
seen and liked it, to find out if it would be a good wintering place for me
and my boat.

It was a disappointment to learn that Bobby Somerset was away, sailing
in the Greek Islands until the autumn. I would have liked to have heard
his ideas about Andraitx and other ports in the Balearics, but I hoped that
I would see him before I stopped my summer sailing.

From Andraitx we sailed north, below the fifty miles of mountain range,
a line of peaks two and three thousand feet high, and then around Cape
Formentor to the wide sandy beaches and flat country of Pollensa and
Alcudia. We found long distances between bays, and we missed the pretty
small anchorages of Ibiza, all so close together. We thought Mallorca a
poor cruising ground in comparison, but we were wrong. We discovered
bays and channels even better than the ones we had loved in Ibiza, on the
opposite side of Mallorca to the coast we first visited. But that was not
until after we had come back from sailing around our next target, the
island of Menorca.

CHAPTER NINE

More Balearics

We let the wind choose whether we went on around Mallorca, or crossed over to the northern island of Menorca. The wind took us to Menorca. It was a change from the other islands, both in scenery and people. Instead of a skyline of sharp high peaks and deep valleys we saw a plateau with a low flat horizon broken by a few separated cones, only one more than a thousand feet high.

The edge of the plateau came very close to the sea, and then became a line of low cliffs and narrow sand or pebble beaches. The cliffs were cut by frequent knife-edge ravines, about a mile apart, so steep and narrow that they were too small for us to use.

This very different island had its own fascinations. It was small, twenty five miles long, and only thirty miles from Mallorca, but it seemed to be in a separate world. Tourists were no novelty to Mallorca and Ibiza, but in Menorca a foreigner was still a strange object. Outside the capital Port Mahon, it was hard to find an interpreter for English or French, and sometimes even for Spanish, which was also a foreign language, far removed from the local version of Catalan.

Life was much more placid and carefree than in the other islands, and the people were more content, not wanting change and modernization.

Sheltered anchorages were not as easy to find as in Ibiza, but those we found had charm and interest. Our first stopping place was a lonely one, almost as deserted as Espalmador. It was Cala Santa Galdana, on the south coast, a round bay embraced by two tall cliffs.

The cliffs were white and grey, crowned with pine trees, some growing from cracks and ledges on the vertical faces. The cliffs nearly overlapped at the entrance and then ran inland, enclosing a beach on one side and a valley on the other.

On the beach, at the foot of the cliffs, were caves with walls and doors, but no windows, built across their mouths. Some fishermen lived in the caves, and had three open boats moored in the curve of the cliff, and a herd of pigs which squealed and rootled along the beach and among the pine trees.

Down the valley flowed a small quiet river. It was the only constant river I saw then or later in the Balearics. Usually the 'rivers' were either dry stony beds in summer, or wild rapids in the winter rains, in keeping with their Spanish names of 'torrentes'. This one was gentle and slow, moving between tall rushes over a muddy bed to enter the bay across a shallow bar of rock and sand.

We hauled the dinghy over the bar and rowed up the river as far as we could go, not much more than a mile. On one side we were bounded by the wooded cliff and its overhanging bushes, and on the other by a marsh of bullrushes. The water was brown, about twenty feet across and a fathom deep, and was full of grey and silver fish, all very frightened of us.

After a mile the river became a stream too narrow for the spread of the oars, so we had to paddle. Finally we were stopped where the stream came out of a cleft in the hillside, by a grove of willow trees and one big eucalyptus. There was a clear spring filling a round pool in a dark cave, and a row of blackberry bushes along a meadow fence, over which a mare and her small foal looked at us. A mile of rowing had taken me a long way in thought from *Leiona* and the sea.

From Galdana we moved right round to the northern coast and found another enclosed harbour, but this time a large one. It was Cala Fornells, like a big lake, spreading two miles inland from a narrow rocky mouth to become a wide bowl bordered by fields and farms.

The village was close inside the entrance, small and clean with shining white houses close together. It gave us the impression of being a smaller

and less extreme version of Ciudad Ibiza. It even had an artist colony as
Ibiza had, but this too was smaller and less extreme.

Fornells was a place of happy memories. We swam and we sailed the
dinghy over the shallow waters among the little islands and creeks around
the edge of the bay. We shopped and had coffee and good meals in the
village, in the cool shade of the thick-walled buildings by day, and by
night on the flat roof-tops under the stars.

Diana was enchanted by the problems of shopping. Visitors were not
expected, and the owners of the small shops took in only just enough
provisions for their normal daily customers. We got used to finding they
could sell us nothing at the time of asking, but would say:

'I am sorry, we have nothing to sell you now, but at eleven o'clock
Jorge will be coming with beans and carrots, and you can buy some
then.'

To buy a simple thing like a chicken took an hour of unsuccessful
searching one morning, until at last a woman waiting in a baker's shop
came to our aid:

'Señora, my cousin has just bought half a chicken, from a house I know
not far from here. Go now with my daughter and she will take you there.
Maybe you will be able to buy the other half.'

Then the baker joined in.

'Yes, and when you have bought the chicken, bring it back here, and I
will cook it for you, with a good Menorcan sauce.'

We found the house and bought the half-chicken. An hour later we
rowed out to the boat with a hot iron dish on our knees, and the scent of
the baker's good sauce in our noses.

Fornells and its wide bay had been a change from the cliffs and caves of
Galdana. Port Mahon was a still greater change, a city at the end of a long
narrow, deep fjord between high cliffs.

Coming to it from the north we could not see the entrance until we had
rounded the massive fortified headland of La Mola. Then suddenly we
were looking between the cliffs of La Mola on the north and San Carlos on
the south, with the long harbour running inland for three miles from a

mouth only two hundred yards wide. On each side of the entrance passage were big iron rings and bollards which had been used for warping through the warships of Nelson's time against the wind. The wind was blowing out of the passage, and we were glad of the engine's help.

The harbour was magnificent. There were steep cliffs to the south and sloping green hillsides to the north. We entered between two shelves of black rock, very close on either side, and then passed the long island of the Lazaretto, where the thick triple walls made the old leper hospital look like another fort. Then we left Isla La Plana to starboard, while to port was the pretty village of Villacarlos, where white and yellow houses grouped themselves in a half-moon around coloured fishing boats like a stage setting.

Then we passed the lighthouse and hospital of Isla del Rey and the bay of Cala Figuera, where we could see four yachts in front of a large white building with a line of palm trees. Then at the end of the bay we were between the docks, ships, and busy launches of the naval base, and the city itself, which lay along the southern cliffs. We could see only the front edge of the city, on its plateau above us, with spires and chimneys and some of the higher roof-tops showing over the faces of the nearest houses. From the centre of the city's front a splendid pair of wide and balustraded stone stairways wound down the cliffs to wide landing steps. At the bottom of the cliff was a single line of houses and storerooms, along a street which was part of a long dock for steamers and fishing boats.

There was no good place for us at this dock, so we turned back to where we had seen the other yachts and lay stern-on in front of the palm trees of the Club Nautico where we were made welcome.

We enjoyed a few days in the city with its staircases and winding streets which opened into angular and sloping 'squares'. Our shortest way to the city's centre was up one of the staircases and then through the covered market which stood beside the big cathedral. We usually managed to get lost, especially when we were coming back with all the day's heavy shopping, and our Spanish improved with the contacts we made by asking for directions.

The people of Port Mahon were eager to stress their historical connections with England, and there were still traces of the British occupation of Menorca, which lasted for nearly the whole of the eighteenth century. In the tavernas we found the Menorcans drinking a locally made, and very good, gin, and some of the windows were of the English sliding pattern instead of the opening type usually met in the Mediterranean.

We heard much about Lord Nelson and Lady Hamilton, although there is evidence that Nelson only stayed in Menorca for a short time, and Lady Hamilton probably not at all. This has not stopped hotels and restaurants being named after them.

There are some other names of interest connected with Menorca, and maybe with better historical reason. They cover a wide range of history, from Mago, Hannibal's brother, for whom Mahon is said to have been named, to mayonnaise sauce, which took its name from the port. Story has it that this sauce was first served at a dinner given in Paris to celebrate the French capture of Menorca from the British. Poor Admiral Byng failed to prevent this capture, and was executed for his failure.

Port Mahon was attractive and interesting, but August had not yet quite ended, and the weather was hot. We wanted to move from our yacht club berth to find cool breezes and to bathe in clean water. We found a good place, at Cala Taulera—Mulberry Cove—behind the big Lazaretto island, in easy shopping reach of Villacarlos.

The cove was a long pool with more than two fathoms in its centre and with two entrances, one too shallow for *Leiona*. We anchored between the low walls of the Lazaretto and the high fortress of La Mola, within sound of its bugle calls.

The water of the pool was dark and green and still. We had it all to ourselves, and found something sinister about our solitude under the silent walls of the empty prison-hospital. The island was deserted. Apart from the birds and grasshoppers and butterflies the only living creature we met on the Lazaretto was a young bay horse. There was something unreal about him. He was strong and well fed, with no mark of saddle, bridle, or traces, and he behaved as if he owned the island. He would come boldly to us to see if we had any food for him, and he would push us aside just as boldly if we happened to block his way on a path. He used to appear from nowhere and pose on the skyline. Then he would jump down the slope like a diver and gallop off out of sight. He could gallop through postern gates which did not look high enough for him, and he could hide himself from our searches until he put in another sudden and striking appearance. There was probably a good explanation for him, but we preferred to leave the mystery unsolved.

He was standing above us in high silhouette, watching, as we left the harbour and sailed for Mallorca.

I was still looking for my winter home. Port Mahon had been a possibility, but I had been told of long periods of rain and strong north winds in the winter. Certainly in our three weeks there we had found the winds much stronger than those we had met in Mallorca and Ibiza, and they had come mainly from the north. Menorca's weather seemed to be closely linked to the unfriendly Gulf of Lions, known for having the worst weather in the western Mediterranean. And even though Port Mahon was in remote Menorca, it was still a city, and I wanted to spend the winter in a village.

I did not find a wintering place during our September of leisurely wandering along the east and south coasts of Mallorca, but I did find what we thought was the best cruising area of all the Balearics. We were glad we had left it to the last.

On both these coasts, and especially on the eastern one, we found what we had missed along the north-west of Mallorca the month before and all around Menorca. There was a greater choice and variety of anchorages, and the scenery was even finer than in Ibiza.

We called in at Cala Retjada, Manacor, and Puerto Colom. The first two were cramped and crowded, and we moved on quickly. At Colom we were comfortably anchored in the middle of a large enclosed harbour where there were two villages sharing the same name, separated by a marshy valley.

Diana was delighted with the pictures she found to paint in the older village, and we stayed until she had finished her work. When at last I persuaded her to let us move I was relieved when we found other places which were as attractive to her, in their different ways, as Colom.

We sailed on south. The wind was light, and our sailing was slow but pleasant. There was enough movement in the air to keep us cool, and the sun and water were warm enough to make bathing a demanding pleasure.

The land we were leaving and picking up was beautiful and varied. At first, in the north-east, the coast was of coloured cliffs and sandy coves and narrow inlets. Then it flattened out into green slopes and later to long beaches and pine-scrub ridges. Always in the background was a range of pointed hills of a thousand feet or more, in line with the coast and five miles inland. On their slopes we could see farms and orchards and lanes, with the monastery of San Salvador overlooking everything from the highest hill.

At Cape Salinas the land turned westward. From there Mallorca's southern coast ran on with sandhills and beaches for fifteen miles and then changed into sheer cliffs as it curved into the wide Palma Bay, ten miles across. After that the shore line took on a background of brown hills and mountains, to meet again at Andraitx and Dragonera the long mountain range of the north-west coast.

In some of the anchorages we stayed a few days, and in some only a few hours. My favourite on the east coast was the little harbour of Puerto Petro with its triple bays. I would have liked to stay there for the winter, but the sheltered inner harbour was not deep enough.

The village at Puerto Petro had no electricity or telephone, or even sources of fresh water. The scenery in the harbour and in the village and in the background was beautiful, and the people were happy and helpful. Our special friends there were Sebastian and Barbara who owned a small white taverna beside the boat pier, hiding behind the broad leaves and mauve flowers of their begambia tree.

Near Puerto Petro was Cala Figuera, looking like a Cornish village, tucked away from the sea at the end of a narrow, forking, cliff-walled channel. We lay at the fishing boat dock on the night of a fiesta, when the

3. My home for one winter, Puerto de Andraitx, Mallorca
4. Palermo, the *Palinuro* and *Leiona*

15. Paolo and Costanza

16. Bonifacio

villagers danced until daylight under the brown nets hauled up on the tall drying posts.

On the south coast we lay for one day in the open anchorage of La Rapita, off a straight sandy beach, and the next evening we hid ourselves in the ravine of Cala Pi, true to its name of 'The Little Bay'. There was only just room enough to turn the boat between the cliffs on either side, twice as high as the mast, and we lay with anchors from bow and stern. We scrambled up the cliff face by a path which was almost a flight of steps, and on our way we met partridges, pigeons, rabbits, and weasels, all more suited to Richard's Dorset than to my Mediterranean.

We went to the big city harbour of Palma, full of ships, yachts and people, and attractive with its old buildings, wide avenues, and twisting streets with courtyards half-seen behind tall gateways. The city looked its best at night, seen from the centre of the harbour. From here we could see the shapes of the buildings and of the old fortress walls, while the coloured lights of the streets and windows drew long wavering lines over the flat black water, and the floodlit cathedral and castle stood out against the sky and the stars.

Still further along the southern coast was the clover-leaf bay of Portales, with its coves and cliffs, a golden oriole, and deep caves which some people said were Phoenician and others said were not. And there was the wide beach of Magaluf, becoming overcrowded with towering hotels and flats, but bringing us to the delightful discovery of Matthew's Bar.

It was a surprise to find Matthew and his bar among the modern hotels of Magaluf. Matthew was simple and old-fashioned, a small, round, cheerful man who was happy with the money he had, living in the house where he had been born. The house looked out of place on the gaudy beach, but pleasingly so. It was hard to see it at all, between two big new hotels, small and hidden among its own trees and vines. Matthew ran it as a combined bar, restaurant, and boarding house with the help of his wife and two waiters. He could have made himself rich by expanding the place or selling the land, but he said he was happy as he was and couldn't see how a change would make him any happier.

The only way we could reach Matthew was by a 'surf landing' in the dinghy. We always got ourselves wet, but it was worth it. Matthew's food was excellent and his company delightful. He used to join us after dinner, bringing a bottle of brandy, the young waiter Antonio with his guitar, and Bobby the performing dog. The evenings lasted until very late, over many cups of coffee, and we were entertained by Bobby's tricks, Antonio's music, and Matthew's stories and songs of Spain and Mallorca.

By now we had nearly completed our circuit of Mallorca, with only the big bays of Santa Ponza and Camp de Mar between us and Andraitx, which had been our starting point. But before we finished the circle we turned off south to visit a group of islands where we spent our happiest week.

These islands were part of Mallorca, but lay five miles south of it, off Cape Salinas. Only one island was inhabited, Isla de Cabrera, the Island of the Goatherd's Wife, and around it lay a dozen smaller islands or rocks. The Spanish fishermen called them all 'Cabrera and her chicks.'

Cabrera was three miles long and three miles wide, five hundred feet high in the centre, and with a deeply indented coastline. It was exactly what I felt a proper island should be, small enough for a man to be able to see all over it from one high point, and yet big enough to fill in a day in sailing around it, and with many choices of anchorages to spend the night.

The entrance into Cabrera's crater-like central bay was hidden by the narrow arms of two long, high, steep peninsulas, leaving an opening not much more than three hundred yards wide. The entrance was overlooked by the ruins of an old castle, hard to distinguish from the pinnacles of rock on which it was built. The bay was a mile long, almost circular, with a long valley leading away from its southern end.

Tucked away below the castle and hidden until seen from well inside the harbour was the village. It was a village of four houses. One house was a taverna owned by Bernardo and his wife Coloma. A family who laid claim to possession of the island lived in one, and Geronimo the fisherman owned another. The fourth and biggest house was the combined military headquarters and home of the commandant of the island, the castle, and the garrison. This was a splendid title for a lieutenant, in charge of fourteen soldiers.

As well as this little group at the village there were four coastguard police and their families, in a long building overlooking the entrance, and a farmer and his family who lived in the valley, beyond the barracks which were big enough for two hundred men. And there was the keeper of the big lighthouse at the south of the island.

Maybe a flock of sheep and the ghosts of a thousand Frenchmen should be added to this census. In a pine wood in the valley there was a half-hidden monument to French prisoners of the Napoleonic wars who had been left on Cabrera to fend for themselves, and who all died there of starvation.

The sheep led a well-organized life, entirely on their own. In the summer they stayed all day in the coolness of a big cave above the village. No one herded them, shut them in, or drove them out. In the evening, when the heat of the sun had gone, the music of their bells would be heard as they filed out of the cave, as Polyphemus's sheep might have done. Then they straggled along the side of the bay to its furthest end where a spring of fresh water kept some stone troughs full. They drank from these and then moved around to the other side of the bay, grazing as they went, to stay there for the night. Early in the morning we would hear their bells again as they returned to drink at the troughs and then go back to the shadow of their cave.

At Cabrera there were good anchorages all round the big bay, or at the village, either just off it or alongside the short dock which was used each Friday by an ancient steamer. There were two other bays, one near the entrance and one on the opposite side of the island.

The nearer bay, Cala Ganduf, had its unique attraction, a blue grotto. It was a beautiful grotto, with a domed white ceiling to absorb and reflect the blues and greens of the water, and with stalactites, stalagmites, and pillars. Richard swam underneath the dinghy when we were inside and his whole body was covered by a cloak of shining blue lights.

On one dark night in the eastern bay, Cala La Olla, I lost Richard over-board, but not for long. We were anchored in a small basin at the head of the bay and had to move in the night when the wind changed and began to blow strongly. We moved to the other side of the bay and anchored again. I was afraid of getting too near the rocks, and anchored too early, when we were still in eight or nine fathoms. In this depth the weight of the chain was much more than Richard was used to dealing with, and it ran out very fast. Before he could check it the whole sixty fathoms of chain had gone. The end was made fast to a ringbolt in the chain locker by a length of nylon rope long enough to come up on to the deck, in case it ever had to be cut free in a hurry. It stopped the end of the chain just forward of the windlass.

Richard took hold of the chain to pull it over the windlass so that he could crank it in. By then a steep sea was running, and *Leiona* was pitching badly, jerking at the chain. Richard started hauling while we were going down a slope, and the chain came in easily. He was taken by surprise when the bow lifted sharply in the dark, and he was pulled forward by the chain. His feet lost their grip, but not his hands, so he went over the bow and under the water. In the darkness, and with the noises of the wind, waves and engine, I saw and heard nothing of all this, but I heard Diana's scream from near the mast. She pointed at the empty water, and then Richard surfaced. We got a rope to him, and then put down the ladder and had him aboard with no damage other than a draught of sea water and a stopped watch.

A chance meeting in Cabrera helped me in my decision about a place for the winter. When we first sailed into the central bay and made for the village we had seen the mast of a yacht of about our size showing over the dock. Two fishing boats were also there, and I thought there would not be room for us as well. We anchored nearby and were getting the dinghy ready when I saw a fair-haired woman and a small girl rowing out to us. They were English, from the yacht at the dock, and they told us that there was just room for us, astern of them.

We moved in to the dock. Our English neighbours were delightful.

They were Tommy and Eryl Ratcliffe, who owned the seventy-year-old cutter *Valmora*. The crew was completed by their small daughter and two large yellow Labrador dogs. If things became too crowded, especially at times of cooking or eating, the dogs were streamed astern in the dinghy. Their boat was beautifully kept, and everything was simple and serviceable. Their life seemed easy and well arranged and they were a very happy family.

That evening they all came to supper with us, except the dogs, who sat watching from the dock. We talked about living in boats, and I told them of my problem to see if they could help. Somewhere in the conversation I said:

'I want to find a place to live in for the winter with my boat, but I haven't been able to decide on one yet. So far, the place that I have liked best is a little harbour on the west side of Mallorca. It has the odd name of Andraitx. Do you know anything about it?'

They both laughed.

'Yes, we do know something about it. That is where we live.'

It seemed too good to be true. They were just the people to answer my questions, as their problems were about the same as mine.

From all that they told me I felt more sure than ever that the little Puerto de Andraitx would be a good place to go for the winter, so I sailed off to have one more look at it before deciding, hoping to be able to ask Bobby Somerset's advice as well.

CHAPTER TEN

Puerto de Andraitx

When I sailed into Puerto de Andraitx for the second time I felt that I was coming home. It was the first harbour which I had entered with knowledge, not finding my way into yet another new one, since leaving St Lucia in the West Indies over eight months ago.

Diana and Richard left me here, and I was alone to think over my possible choice of a winter anchorage.

In the middle of the harbour a big yacht was anchored, a white yawl. She was *Thanet*. Bobby Somerset was back from Greece. I went up to his house to see him. He was very helpful, and everything he said confirmed what I had been told about Andraitx. I decided to stay.

The village was a group of white and pale brown houses stretching from the end of the bay to the inner breakwater. There were a few houses outside the village, some along the shores of the bay, some among the trees of the hills on either side.

Two or three miles away, beyond the orchards of the valley which led to the bay, was a cluster of brown houses and a big church. I learned that this was the village proper, El Pueblo de Andraitx, while I was at its harbour, El Puerto. I was told that this splitting into two parts came from the days of the pirates and corsairs who used to arrive at a harbour or beach by night in their silent boats and then burst upon a sleeping populace. To protect themselves against surprise attacks, the fishermen and farmers in many Mediterranean coastal parts had withdrawn themselves, their families and their animals a few miles inland, usually to a hill on which the village could huddle within the safety of its walls.

They had to leave their boats on the beach, and some men would be left there to look after them and also to give an alarm if any raiders should come. Sometimes, to give these watchers a better view and some protection, towers were built overlooking the way that the pirates might use for their approach. These towers are found all around the Balearics, and there was a big one at the Puerto de Andraitx, on La Mola, the high peninsula which hid the harbour from the south. As peace began to spread over the Mediterranean people moved back to the sea, but the tradition has not died, and I found many who fished or worked in the Puerto still had their homes in the Pueblo.

The Puerto had great appeal, especially for someone with a boat. There was a sheltered anchorage at the outer breakwater, but the price of its peace and privacy was a row of half a mile, or a walk of a full one, to reach the shops. Near the village there was the choice of going stern-on to the quay, or of anchoring in the rather limited space of the inner harbour. I chose to go to the quay, along the breakwater. This was the place for yachts, while the wide quay in front of the village was kept for the visiting trawlers and the trading schooners when they came for loading or shelter. The small local boats lay beyond the schooner quay, and beyond them again was a simple wooden slipway with a hand capstan at its head.

In spite of its smallness the port seemed to provide what I wanted. I found three butchers, three grocers, two mechanics, four tavernas, two barbers, three restaurants, and a post office. There was no cinema and no petrol pump, and although there was a telephone exchange there were hardly any telephones for it to control. I didn't think these shortages would worry me. I was selfishly pleased to find no bathing beach and no luxury hotels, so that the harbour was free from the noise of speedboats and water skiers. I saw English, French, and Americans in yachts and ashore, and the port was full of people and cars. I did not realize how different and pleasantly quiet it would be when the summer was over.

Soon after my arrival, my Cabrera friends, the Ratcliffe family, came back to their home port and brought a surprise for me. They were going to spend the winter in France, to give their daughter a more sophisticated education than Andraitx could provide, and were leaving their boat laid up and their house empty. They offered me their house at a very low rent, and I was glad to take it. So I found myself with a harbour for the winter and a house from which I could see my boat across the bay.

I was glad that *Leiona* was in sight, because when people learned I was staying they told me terrible tales of the winter storms and of the damage done in the harbour. Nearly every yacht had its story of dragged anchors, broken moorings, or parted warps. At first I felt that I had come to a place of sure destruction, but when I thought it over I realized that the others had survived, and so my boat probably could, too.

I was happy at Andraitx. The little house stood in sunlight all day long, and there were few days without sun. The cheerful maid, Antonia, came to work and sing for two hours each day, except when she went off to Ibiza to visit her family for a week and came back two months later, while Anita took her place, with less singing but with more work.

It was a delight to start the autumn days with a breakfast of fresh bread and honey, figs picked from the tree above my head, and a view across the water of the boats, the houses and the mountains.

The only trouble with the house was the distraction it brought, either of things to look at or of people to talk with. I was trying to work at writing, but I found it hard. There was always something interesting to watch. If there were not boats moving in the harbour there were gulls circling or swimming, or a pair of ospreys soaring over the shallows and then swooping onto a fish, to take it with a splash. Or else a flight of goldfinches, red, yellow and brown as they flickered from tree to bush, or the black and white flash of a hoopoe's barred wings crossing the stream-bed.

The house seemed to attract visitors. Pepe the gardener worked with his mule in the two fields between me and the shore. He was always ready to give the mule a rest and to come up to the terrace with a hat full of tomatoes or green peppers. Then he would settle for a while, sitting on the warm stone wall with a glass of wine or brandy, and give me news and advice. Most of this was wasted on me, but as time went on I began to understand Spanish better, or at least the Spanish which the fishermen and villagers used with me. I used to carry a small dictionary in my pocket so that I could do better than just point when I went shopping.

A dusty track led from the shore road past the house, close along one side and then along the back wall on its way to the farms and houses higher up the slope. Down this track would come Jorge the farmer in his high-wheeled trap, or Juan the mechanic, hurrying on his motor-cycle to mend the water pump of some desperate foreign resident. They would find time to stay for a drink and a cigarette and talk about life, and I learned more Spanish but did less work.

It was hard to do things quickly. The ten-minute walk into the port was apt to turn into a half-hour one, because there was always someone to meet on the road, and a quick visit to buy two things could easily spread to an hour, moving between two shops. The people of the port were friendly and interesting, and so were the foreigners who lived there. Most of these were English who had come to retire in a place of better climate and cheaper living than their own country. There were about twenty English, a few Americans, and two or three each Dutch, Swedish, French, Danes, and Germans. Some had come to stay, some to visit, and some to write or paint. Six of the English owned yachts, all small except for *Thanet*, but only two couples were living on board their boats.

My first problem was to get my boat on to a slipway for anti-fouling. This brought me into contact with the Vera family, on whom the whole village seemed to be based. In the port it was hard not to be a Vera either by blood or by marriage, because there were so many of them. When I first came I had seen a small red trawler at the quay with on her bow the strange name of '15 Hnos'. I discovered that this was an abbreviation for 'Quince Hermanos' . . . 'Fifteen Brothers'. She was the Veras' boat, and I learned that there really were fifteen brothers, or rather fifteen brothers and sisters, but mostly brothers. All except one were alive, and as they were in their fifties and sixties, with children, grandchildren, and maybe some great-grandchildren, they did much towards filling a small village.

The two of the older generation with whom I had the most contact were Bartolo and Juanito. Bartolo was the head of the clan. He gave the orders to the red trawler and was the spokesman for the slipway. He could usually be found mending the nets spread out on the quay, a small thick-set man, with short legs, broad shoulders, glasses, and a keen and cheerful look. Juanito helped at the slipway, and he also owned a little lateen-sailed open boat. He was taller than Bartolo, but had the same strong body and rolling walk of all the Veras. He loved sailing. All through the year the two raking sails of his *Nereida* could be seen moving across the harbour or far out to sea. Sometimes his voice could be heard in Caruso-like song, and occasionally there was the strange sight of Juanito standing on his head, braced on his elbows with his legs in the air while *Nereida* sailed herself.

Juanito slept very little. When he was not sailing by day he was fishing by night, and yet he found time to paint weird pictures in strong colours. These pictures were of ships and sea-creatures, in strange whirling shapes, sometimes showing bombs and torpedoes, and were Juanito's expression on some world problem of politics or religion.

Bartolo Vera arranged a date to slip *Leiona*, and that was my introduction to how the Spaniard prefers working from crisis to crisis rather than to a well-arranged plan. We had decided on a Wednesday for the slipping, two days after the Ratcliffes' departure to France, giving me time to get settled in the house before the boat came out of the water. Late on Sunday evening Bartolo met me by chance on the waterfront and said that everything would be ready for slipping early the next morning. I was not nearly ready, but, like most things to do with boats, the difficulties were not impossible.

It took me a long time to understand this dislike for a long-term plan. When I took a job to be done to one of the carpenters or mechanics I used to make the mistake of saying that it was not urgent, but that I would like it finished by the end of the month. The end of the month would come, and so would the end of the next one, but more urgent requests would get in the way of my less pressing bid. However, in time my job would become urgent, too, and at last by demanding it for the next day, or that night, I would get it done.

Hauling the boat out of the water, and putting her back again, was an adventure. My only experience of slipping had been at Grant's yard in Martinique. There, the slipway was of iron rails on a concrete bed, with a wheeled cradle pulled up by a diesel winch. In Andraitx it was a very different sort of thing. The cradle was like a big sled, with two thick wooden walls, shaped to hold the hulls of fishing boats like the red *Quince Hermanos*, held together by curved iron bars on which the keel would rest. This cradle slid on heavy wooden sleepers pinned across two thick wooden rails. There was no cement below the rails which lay, apparently held by their weight alone, on the sand of the beach. They were not particularly level, and I could see nothing to stop the cradle slipping sideways. However, the system had been working for a long time, so I supposed it would be all right for me.

The cradle was hauled up by a cable leading to a winch with two handles. Beyond the winch was a big capstan with a horizontal beam for turning it. I did not see why there were two hauling machines, but I soon found out.

Early on the Monday morning I moved to the slipway and found Juanito already there with a work boat in which were wedges, floats, and hooks. The cradle had been let into the water, and was now out of sight, except for the heads of some stakes. We put out a stern line to a buoy and a bowline to the winch, and then centred *Leiona* over the cradle. Juanito was in a pair of bathing shorts, and had even taken off his cap, showing a bald head in white contrast to the rest of his tan. He hooked *Leiona* to the cradle and we began to move her slowly forward with the bowline while Bartolo and his men on the shore winched in the cradle.

As the cradle rose on the sloping rails the tops of its side walls began to make contact with the boat's hull and the iron bars touched the keel. Juanito was in the water by now, placing and driving in wedges. Soon the full weight of the boat and the cradle came on the rails, and on the winch. Then there was a delay. This was where the capstan came into play, and for the capstan a mule was needed. In the change of day from Wednesday to Monday there had been a muddle over the order for the mule, and he and his owner were working somewhere up the hillside. Eventually they arrived and the slipping was completed.

As the boat moved up the rails she began to lean over to starboard and I nervously pointed this out, asking if it was all right. I was assured:

'Oh, yes, it is quite all right. There is no need to be disturbed. It is only because the rails are bent. In a moment you will be inclined the other way.' We did come straight again, and then took a list to port which seemed even worse. At last some wedges were added on the port side, and we stood almost level.

There was not much work to be done on the boat once she was out of the water. We scrubbed and washed the hull, looked at the propeller shaft, and put new packing in the stern gland. Then we had two days of painting on the anti-fouling and were ready to go back into the water.

Coming out had been a slow process. Going back was a quick one, once things had started moving. At first the boat would not move at all, in spite of efforts with crowbars and sledge-hammers. In Martinique the launching had been a very orderly process, the wheeled trolley being let gently into the water and even held for a while half-way down while the boat's engine was started and run.

This did not happen at Andraitx. The only control applied was one of my long warps made fast near the capstan with the coils on my foredeck. My job, I was told, was to stand by the coils and use the warp to check the boat's speed as she slid into the water. A check was needed, because astern was shallow water and sticky mud, with rocks from the road embankment on one side and a big trawler on the other, lying to a stern-line from the shore.

To start the launching, the cable and chains from the winch and capstan were loosened and then freed. Only gravity and friction were now holding the cradle on the wooden sleepers. These and the runners of the cradle had been rubbed with tallow and grease, but the parts in contact were still dry and held the weight of the boat, in spite of the slope of the runway.

Bartolo and his men tried to get the cradle moving by raising it with wedges and shifting it from side to side with crowbars, but it only slid a few inches and then stuck again. Someone had the idea of unhitching the shore end of my warp from the capstan and making it fast to a post near *Leiona*'s stern, so that I could help by pulling from the bow. This was not a good idea. I don't think my effort added much to the effort of the others, and I lost what little control I might have had when the warp was leading forward from the bow. I felt exposed and lonely on board, high above the ground.

The heaving and the hammering went on without any good result. At one time the cable was hitched on again and the cradle hauled forward to let more tallow be smeared on the ways. I began to wonder if the delay would go on so long that the new and expensive anti-fouling paint would exceed its time limit and become dry and useless. Suddenly the boat moved and rushed down the slip. There was nothing orderly about this launching. Wedges flew, sleepers rattled, and men jumped out of the way. I had no time to take up the slack of my bow warp before the coils started snaking away through the fairlead.

I grabbed at the thick rope, had it torn out of my hands, and grabbed again. I let go at once, with burning hands, and saw that in the first grab I had lost not only the rope but also the skin from my palms and fingers. The boat shot across the water, heading for the rocks. Luckily the stern-line of the big trawler was slack and low in the water, so that we hit it with our keel and we slowed down without damage. The next time *Leiona* was put in the water I made sure that the coils of the bow warp were on the shore, and that the line was passed around a bollard and handled by six men.

After this excitement I settled down for a quiet winter. I found it very pleasant. There was writing to do and work on the boat, there were letters to answer, and books to read, and many people to meet. In the autumn I had been impatient for the coming of spring and its sailing. Now I found there was plenty to do in the winter, which began to end too quickly.

Work on board *Leiona* took up most of my time. It was all new to me, and I was slow at it, usually having to ask advice on what to do and where to buy what I needed. During the summer's sailing I had found nothing wrong with the boat, but there were many little things to do before the winter really set in, and later some small changes to add to the comfort and convenience aboard. Usually alone, but sometimes with one of the fishermen to help, I put on some more deck paint and varnish to guard against wind and rain, and rubbed the wire rigging with linseed oil. Winches, rigging screws, and steering gear needed greasing and covering, and blocks and shackles were oiled and greased and put away. Then I was ready for whatever bad weather the winter would bring.

CHAPTER ELEVEN

Mallorcan Winter

My winter days began to find a pattern. After breakfast, outside in the sun, I would write for a while until Anita or Antonia came to start the housework, and then go down to the port. Work in the boat would go on into the afternoon, or maybe finish around midday. Anyhow, noon was the time when the mail and papers arrived, and it was always a temptation to go to get them.

The collection of mail at Puerto de Andraitx was a very special proceeding. It was no good going to the post office. This was a hard place to find, following a winding route with occasional signs, rather like a treasure hunt. Finding it did not help, because the postmistress preferred to give out the mail in one of the tavernas. She was an impressive lady named Masiana, dressed always in black trimmed with what the Victorians called bugles and jets.

She used to arrive at the chosen taverna twice a day, at noon and again at sunset, seated majestically in the front of an aged bus, driven by Baltasar, her husband. There was an air of expectancy in the village before the bus came, and a look of triumph among its passengers when they arrived, as if they had completed a long and dangerous journey, and not just the three miles from the Pueblo. The vehicle was narrow and upright, with a high luggage rack on its roof. The whole effect was that of a stage coach reaching safety after defeating a Red Indian ambush.

Señora Masiana would dismount and move with solemnity to a table at the far end of the long room. From her black bag she would bring out small bundles of mail, undo them, reading each address carefully, and then in silence begin to spread the letters out in front of her. While this was going on, a dozen or more hopeful clients, Spanish and foreign, would be crowding the table, watching for letters, but not daring to speak. An even worse crime than asking and interrupting the process of selection was to take a letter. One Englishman did so, and in spite of his three-score years and ten, was slapped on the hand like a naughty schoolboy and was given his mail last of all.

Another Englishman was sure that some letters had been sent to him, but nothing came. At last he persuaded Masiana to let him join her in the search. He found his letters in a neat pile in a drawer of her desk. When he began to complain about her carelessness he was silenced by the unabashed reply:

'Of course I knew the letters were there, but how could I know they were for you? Look, some of the words are very small and badly written. From all that I could see it was apparent that these letters were for some Señor Esq. I did not know any Señor Esq., and he had not come to ask for his letters.'

During the winter I began to speak Spanish with some fluency, if not correctly. Most of my conversations were with fishermen and farmers, and words of Mallorquin dialect were apt to creep in. Mallorquin is an offshoot from Catalan, and is more truly a French dialect than a Spanish one.

One morning the baker was astonished when I confused *cansado* with *casado*, and told him that he looked 'very married' when I meant 'very tired'. And once I was slightly shocked by a Spanish lady's reply to my conventional polite inquiry after her health. She said she was '*muy costipada*', which I thought was too intimate a revelation between strangers until I learned that all she was describing was a cold in the head.

Most of my dealings with the villagers were to do with *Leiona*, and so my main contacts were with the Vera family who were always ready to give me help or good advice. I also used to go to three carpenters, to Juan for straightforward simple work, to Tomeo for shipwright tasks, and to Pepe, his nephew, for jobs needing mechanical tools. I discovered the different fields of the two mechanics, that Juan was the man for blacksmith work and petrol engines, and Miguel for electrical and diesel repairs. For a long time I had been wary of going to Miguel for anything, because the only tools I ever saw him carrying were a five-pound hammer and an eighteen-inch spanner. He explained to me that these were only for use when all else failed and that he hoped to find the more normal tools on board.

The carpentry work in the boat was only for minor changes, but it took a long time to organize and complete. The boom gallows which Mr Gilbert had made in Antigua had proved itself a good idea, and so had his 'bridge' across the cockpit floor. Tomeo the carpenter made a refinement to this for me. Out of a block of hardwood he carved a shape like a loaf of bread with four small feet. These feet fitted into the holes of Mr Gilbert's grating on the bridge and allowed the block of wood to become a movable foot-rest for any length of leg, a great comfort when the boat was heeled.

A large wooden combined tool box and work bench was made, portable and kept lashed against the starboard bunk in the saloon. With a cushion to fit the top it made a good seat for that side of the swinging table and did away with the long stretch from the bunk. Four bookshelves and two magazine racks helped to keep enough reading on board. The old direction-finding wireless set was replaced by a small one, and this left room in the formerly cramped chart table so that charts could be used without being folded. And I sold the radio transmitter, which I had not used since my conversations with Vernon Nicholson when leaving the West Indies. The buyer was on his way to the Caribbean and his first wireless contact was with Vernon.

Many of the small refinements gradually became a part of the boat until, like Murlo's knife drawer, it was hard to remember that they had not always been there. A waterproof box for matches was a help, fixed to a galley shelf, and so was the little trough of water below it in which to drop the used matches and avoid burned fingers or, much worse, holes in the thick white rubber flooring which had replaced the old fitted carpet and cracking linoleum.

A bigger change was the reeving of Terylene halyards for both the headsails instead of the old galvanized wire ropes. Open-barrelled halyard winches took the place of the self-stowing types, and I found them easier, quicker and safer to use even though there were now ropes to coil.

I had a gang-plank made for using when we moored in the Mediterranean style with the stern to a quay. Its production was a lengthy process. To work out its size and detail three men had to meet and confer on board *Leiona*, Tomeo the carpenter, Juan the mechanic, and myself. This meeting was hard to arrange. When I could get hold of Tomeo, Juan would be away in Palma or working in someone's house. If the day when I could find them both was a windy one Juan would not come on board, because the slightest movement of a boat made him feel sick. On a fine day when I could get Juan, Tomeo would be in another village mending a roof. At last a very effective plank was made, but the time was spread over three months.

The most efficient people in the village were probably Margarita of the telephones and Margarita of the shop. Margarita of the shop owned the one haberdashery in the village, of much greater scope than just cloth and threads. She knew the best seamstresses, and it was to her I went for repairs of clothing, and to have curtains, sailing clothes, or seat covers made. Her husband drove one of the two taxis of the village, and through him Margarita could arrange for the developing of films or filling of gas bottles, or anything else where recourse to Palma was needed. Demand had led to supply, and so Margarita had a stock of nearly every small thing that might be wanted, even to shoes and sealing wax, if not ships.

Margarita of the telephones knew how to find everybody and everything. There were not many telephones on her switchboard, but she was kept very busy because her calls involved booking passages, sending telegrams and messages, and coping with long-distance calls to other countries for foreigners. She did all this in Spanish or French and spoke no English at all, although I am sure she had a hidden knowledge of it. As well as controlling the telephone exchange and its complications she was a go-between for a variety of things. If you wanted a cart of firewood or a truck-load of water, Margarita was the person to ask. She would deal with soda siphons, dry cleaning, newspapers, or doctors' prescriptions. She would find you a plumber, an electrician, or a taxi.

She had one weakness, an enthusiasm for picking wild mushrooms. For a week in the autumn a local type of mushroom made a brief appearance on the hillsides. During this week Margarita was away in the hills all day, leaving her husband and son to cope with the telephones and all the rest. They were willing but ineffective substitutes and the foreign community was glad when the mushroom season was over.

The winter gales did come, but they were not as bad as I had expected. An early, and very rainy, gale in October put one small yacht on to the rocks, but only because she had been anchored badly in an exposed cove. There was a short gale in December, and a patch of four bad days in February, but otherwise there were no winds of danger or great discomfort. After Christmas we had the traditional *Calmas de enero*, the Calms of January, when for three weeks the summer seemed to have come back.

There was usually something of interest to see in the port. The processions of the holy days brought pleasure and sympathy to those who watched as well as to those taking part. There was the solemn and eerie procession to the cemetery on All Souls' Night, with candles in the darkness against black clothes and veils. There was the sad procession of Good Friday and the happy one of Easter. The two events which I most enjoyed were the Blessing of the Animals on St Anthony's Day and the Coming of the Three Kings at Twelfth Night.

The Blessing of the Animals was a happy affair, with no mysteries. Early in the afternoon people started moving up to the church to gather below the steps. Some of them were leading or carrying animals. There were dogs but no cats; rabbits, pigeons, canaries and goldfinches in cages, a pig, a donkey, a foal, and even a tortoise. The farm carts and their horses were decorated, and many of the children were in fancy dress or local costume.

After a while, when the excitements of meeting were over and the dog fights quelled, the priest came from the church with an escort of choir boys in scarlet. He said a prayer and gave a short address. Then the animals were brought past him, and each one was sprinkled with holy water. Each owner gave one of the choir boys a coin and was given a printed certificate of the blessing of his animal.

When all the blessings had been done the ceremony turned into a dance on the wide platform above the steps and in front of the church door. The onlookers crowded around the platform, leaving a small space at one side of the church door. Pepe the farmer was already seated there with his guitar and he began to play. Three girls came forward, laughing back at the joking comments of their friends. They danced a local dance well, and had obviously rehearsed it. They were followed by an eightsome of very young children who hopped around shyly, forgetting the steps but receiving loud applause from the crowd and encouragement from their mothers.

Then the dancing was thrown open for anyone who wanted to try, or who was called out by popular demand. Some started boldly but soon lost the pattern of the steps and ran off embarrassed. The most elegant dancer was one of the older Vera brothers, wearing a green golf cap, rotund and red-faced with his equally rotund wife. They danced with such grace that it was easy to imagine powdered wigs and lace cuffs.

Youth had to have its fling, and after an hour or more of the local dancing the modern dances took over with their twists and shakes. Unlike the peasant dances, these seemed out of setting under the shadow of the church.

The Coming of the Three Kings was the great event of the winter. Spanish children do not have their presents brought to them on Christmas Eve by St Nicholas, but on the eve of the 6th of January, Twelfth Night, by the Three Kings, the Three Wise Men who brought their gifts of gold and frankincense and myrrh to Bethlehem.

In Puerto de Andraitx the Kings arrived in a fitting way, by boat. All day long the children were excited and expectant, and by sunset this feeling had spread to their parents and the other grown-ups. As it grew dark everyone gathered towards the dock in front of the square. Big Mateo trundled a hand cart from the direction of his fishery storehouse

and from it unloaded long tar-soaked torches. Three saddled and capari-
soned horses were led to the dock, clattering on the stone paving and
dancing about in their nervousness of the crowd. The two taxis of the port
drove down to point their lights at the landing place, and three carts
decorated with palm fronds moved into a dark corner of the square.

The crowd packed more closely, pushing forward to the edge of the
dock, leaving an empty space near Mateo and his torches and a narrow
corridor from it into the square. The laughter and talking died away.
Everyone looked towards the lighthouse, a black finger flashing against
the almost faded glow of the sunset. Then a red rocket climbed up above
the flash of the light, drawing from the crowd the rising, sighing cry that
always follows rockets.

Mateo lit his torches and gave them out to a dozen small boys who
formed a half-circle at the dockside. The horses threw their heads and
drew away from the flames. Then a red and orange glow was seen over the
line of the inner breakwater, moving in from the west. The glow spread
past the end of the breakwater and a boat came in sight, turning in towards
the crowd on the dock, a fountain of waving torches and coloured lights.
Sparks floated up in the darkness. The flames and lamps shone against the
sides of the boat and were reflected in the water.

The children shrieked with joy as the boat touched the dock, and everyone waved and shouted. The cars switched their lights on and off and blew
their horns. The Three Kings stood proudly in the bow, in golden crowns
and purple robes. King Baltasar's face was blackened. All three stepped
ashore in silence and dignity and strode to the waiting horses. They
mounted and rode up the hill to the church, flanked by the torchbearers
and followed by the crowd. The tall doors of the church were opened
wide and made a pointed arch of brilliant light against the dark of the
trees and houses. The Kings left their horses with the link-boys and went
inside to present their offerings and receive a blessing.

The carts had been driven to the church steps and each stood behind
one of the horses. In the light from the door the loads of parcels could be
seen, wrapped in gay paper and in charge of laughing young men and girls

in their best clothes. The Kings came out of the church and rode away into the village streets, each in a different direction, followed by his own cart of presents.

They stopped at a few houses to give out presents, but the normal stops were at crossroads. There the King would call out the names of the children of that neighbourhood for whom his cart held presents. There was a typically delightful Puerto confusion at the first stop of all. This was at the house of the Harbour Master, an important stop, not only because of his appointment, but also because he had nine children. King Baltasar's herald knocked loudly on the door and called out, 'Come to the door! The Kings are here.' There was no answer. The house was empty and in darkness. A message was passed back through the crowd and all the family came hurrying to their home to open and enter it so that they could receive the presents from the King and give the herald his reward.

The square and the streets emptied slowly as the children and their parents went home to open the parcels and see what the Kings had brought them. For the next few days the children played with their presents and we could see what the gifts had been. The girls had domestic presents of dolls and cradles and cooking sets, but the boys nearly all became warriors of some sort. Silver-plated revolvers in studded twin holsters were a favourite equipment, but a wide span of history was covered, from the Siege of Troy to the Space Age.

As the days passed there were exchanges of weapons and clothing, and history became confused. It was surprising but not uncommon to see a Roman legionary repelling with Frontier Colts the attack of two Confederate cavalrymen armed with a javelin and a death-ray gun.

If my day's work had finished early enough I would go for a walk in the hills, where the mule-tracks ran along the slopes and through the woods. The views were wonderful, down into the valleys, or along the coast, or out to sea as far as Ibiza on a clear day.

Before Christmas the soil in the fields and orchards was either bright red or dark brown from the new ploughing, or fresh green with the shoots of new barley or beans. Oranges and lemons were shining on their dark-leaved trees, and there was a look of pinkness where the buds were swelling on the almond branches.

The mimosa was already flowering, and the big round trees were hanging their feathery clusters of pale yellow boughs over the walls of the gardens and courtyards. Wild flowers were spreading along the road-sides and in the meadows.

With the new year came the flowering of the almond blossom, when all the valleys and hillsides turned to whiteness with scattered patches of pink.

As the spring came the fig trees put out pale green shoots on the thick fingers of their bare branches, and the green of the fields was almost hidden by the blues and whites and yellows of the flowers. The almond petals were blown away to lie on the ground like snow, and the orchards changed to the colour of new leaf.

When April came, and it was time to start sailing again, the hillsides were all green and there were red carpets of poppies below the fig and carob trees.

Early in the spring I put *Leiona* on the Veras' slipway again for new anti-fouling and for painting the topsides. I got a fisherman who lived in the Pueblo, Bartolomeo Ginart, to help me with the work. After launching there was still a lot to do, in varnishing the mast, oiling the rigging, and painting the deck. I asked Bartolomeo if he knew of another man whom I could engage for the next couple of weeks. He thought for a while.

'Yes, Señor, there are two or three, but I have another idea. My wife knows how to work on a yacht, inside and out. She works twice as hard as a man, and you would only have to pay her half as much. What do you think about it?'

I thought it a good idea.

Next morning Josefa came to work, a happy, wide-shouldered woman with her hair in a pony-tail. She was a great success. She was even more of a singer than Bartolomeo, and her tunes were more cheerful. She was a hard worker and was strong enough to pull her husband up the mast in the bosun's chair. This left me free for other jobs and to go ashore to create some crisis that would ensure my requests being remembered.

We had good weather for the work, and got through it quickly, anchored peacefully in the middle of the harbour away from the dust and distractions of the quay. My helpers used to arrive early in the morning and I would collect them in the dinghy. They brought their own basket of lunch, and at noon I would put out a flagon of wine and go off to sail the dinghy for two hours while they had their food and siesta. Sometimes they brought their pretty young daughter Carmen as an unpaid helper, but she was of decorative value only, and put in more time at siesta than at sandpapering.

By early April everything was done, and I was ready to sail. My winter in Mallorca had been a good one. I was sorry to leave Andraitx and the Balearics, but I wanted to see more of the Mediterranean. I had only just entered it, and there was a lot more to cover.

My plan for the coming summer was to go as far as Malta and then to see as much as I could of Sicily, Sardinia, and the coast and islands along the west of Italy.

CHAPTER TWELVE

Eastward to Malta

For this summer of sailing in the Italian Sea I had old friends as crew. Richard and Diana could join me in Sicily in late May, and Michel and Marie-Claire wanted to come for a month, but would have to leave before the end of May. This seemed a perfect arrangement, but it did not work out quite so easily.

My French crew arrived in April to stay in my house and help me get *Leiona* ready. They brought bad news with them. Marie-Claire's father had refused to let her sail with us.

It seemed odd that he should have agreed to letting his daughter sail across the Atlantic the year before, and now forbade her the almost coastal voyage we were planning in the Mediterranean. I think that he had been surprised and forced into agreeing to the Atlantic trip before he had time to think. This time he was adamant. He said crossing the Atlantic was bad enough, but braving the Mediterranean before June was much more dangerous. Having heard some of the stories about it I was beginning to agree with him.

Anyhow, Marie-Claire could not come.

Michel was seldom at a loss and he had a solution ready. He knew of a young Frenchman in Paris who could come at once if I wanted him. Michel thought he knew nothing about sailing but had some experience of small motor-boats, and was not likely to be seasick. We sent him a cable, and Michel went off to take Marie-Claire to France. Four days later he came back with Pierre Dubois.

Pierre looked lost when he stepped on board. Michel had been wrong about his small boat experience. He had only looked at the sea from the decks of liners, and he was obviously not happy about being so close to it. I had hoped that before starting our journey we would be able to do some gentle short sailing to get used to a boat after months ashore, but time and weather were against us.

Our start was delayed. April opened with some spells of bad weather and then with a long stretch of strong east winds. We were going eastward so we waited for the wind to change from blowing straight in our teeth. We had to wait until nearly the end of April.

I had made my farewells in Andraitx, but with reservations, as I did not know where I wanted to stay for the coming winter. Unless I found a place I liked better in the Italian waters I was ready to come back to Andraitx at the end of the summer. This would add distance to my plan for the following summer, sailing in the Greek Islands, but I could find time to cover it. Bobby Somerset said I would surely be back.

We made two stops along the Mallorcan coast and were held up once more by bad weather. Then we set off for the southern end of Sardinia. My crew were not much help at the start. Poor Pierre was very seasick for the first day. Michel developed a sore throat and a painful earache, and on the second day began to have a fever and a high temperature. I thought of turning back to let a doctor have a look at him, but by then we were as near to Sardinia as to Mallorca, and the wind was with us, so we went on.

Soon everything got better except Michel's fever. The wind eased and the sea flattened. We moved steadily eastward at four knots under a warm sun, and Pierre got over his seasickness. He was still looking at the whole affair with some doubt and was worried by the empty sea and unbroken horizon. He did little talking but a lot of thinking. He had trouble at first over steering a steady compass course, and I tried to help him by saying:

'Pierre, don't try to make the compass move about. The compass doesn't move; the boat does. That black line shows where the boat is pointing. Make it move to the number you want on the compass by moving the wheel the same way.'

He had less difficulty after that, but he was very silent. The next morning he said to me:

'Keith, what you told me about the compass not moving was very interesting and has been very useful. But I think maybe it is not correct?'

On our first day in the open sea he had watched me take sights and work out a position, but had said nothing. On the second day, with still no land to be seen, he could not contain himself any longer. Trying hard, but failing badly, to be lightly casual he asked:

'Keith, have you an idea where we are?'

I should have been more definite in my answer and given him confidence, but I said:

'Yes, Pierre, I have an idea, but of course it may not be right. I hope it is.'

'Oh . . . Yes . . . I see . . . When do you think you will know certainly?'

'Well, this afternoon we'll alter course and steer fifteen degrees more to the south. If we are in the right place now, we ought to see the light from San Pietro about midnight.'

Michel was still not well enough to take a watch and I was sharing the work with Pierre, he steering for two hours and I taking over for four. I handed over to him at midnight. I had hoped to be able to show him the San Pietro light by then and remove some of his worries, but there was no sign of it. That was probably a good thing because half an hour later he woke me with an excited cry of 'I can see a light! I can see a light!'

I could hear the relief and happiness in his voice. To have seen it himself, at the right time and in the right place, made him feel at last that we could find our way around the sea. He lost his worried look and from then on enjoyed the sailing.

Shortly before dawn the moon set as a deep red triangle. A pale pink sunrise outlined the Sardinian coast for us and the day spread light over it. We sailed southward, following the line of the shore and heading for Cape Teulada, the southern point of Sardinia, where the chart showed a small harbour. Michel was much better by now, but it seemed a good idea to let a doctor see him and stock him up with medicines. We could afford the time for a short stop, and a quiet night's sleep at anchor would do us all good.

It was a good day's sailing, bright and hot, with a light wind to take us over a calm sea. The shore was pretty, low ridges of hills with steep white and orange cliffs. The fields and small woods on the slopes still had their springtime greenness, and there were clumps of bushes like gorse in yellow flower. White houses were scattered about in the green, but we saw no towns or villages. In the afternoon we passed the Cape and rounded up to anchor in the shelter of Teulada's breakwater.

We were glad to have a night of rest, and really would have liked to have stayed longer. The small curving breakwater formed a good harbour, empty except for one deserted motor boat and two naval launches. Further up the bay, in shallow water, was the fishing port and the village. Around us were rocky hills above sandy coves. Sheep and cattle grazed on the short turf between the gorse bushes, and the sound of their bells came to us in the quiet evening. Michel and Pierre rowed ashore to go to the village. They came back carrying fresh fruit and bread. Michel had not found a doctor, but was obviously feeling much better.

My two Frenchmen asked if we could make our next stop somewhere in Sicily, preferably at Palermo where Pierre had Sicilian cousins whom he wanted to see. I would have liked to agree, but I felt that we had to get on to Malta. I had ordered stores and gear from England to meet us there, and I knew that it was all waiting for collection. Also, I wanted to see Malta and spend some time there before sailing to meet Richard and Diana at Augusta, on the east coast of Sicily. Pierre could do his Palermo visiting after that. So we decided to sail straight for Malta, passing south of Sicily.

As it happened, we did visit Sicily before Malta. After leaving Sardinia we had a good fast sail until we were near Marittimo, the most western of the Aegadean Isles which lie just off the Sicilian west coast. I was glad when we saw its light and knew we were safely past the one hidden obstruction we could meet on our way. This was a sunken rock, only just under water, fifty miles from any land and twenty miles south of our course to Marittimo. I had deliberately plotted a slight 'jink' to bring us close to Marittimo as a check, because the direct route from Teulada to Malta passed only a mile from the rock. Even the name of this rock bore a sinister coincidence. I had seen it on the chart when planning the journey during the winter, and I had been a little startled to see that it was called 'Keith Reef'.

After we reached Marittimo, the wind began to come from different directions. In the dark of the morning we were hit by a sharp squall from the north-east. This blew for about an hour and then died away completely. An hour later it began to blow just as hard again, but this time from the north-west. Then the sun brought with it a light, southerly wind, a calm sea and a hot morning. We went along smoothly and slowly until the afternoon, when a big swell started to come from the south-east, from straight ahead. It increased very quickly and was followed in three hours by the wind. Before sunset we were being knocked about by a sirocco, not quite gale force, but very uncomfortable.

In the Mediterranean bad weather often comes with little warning, but it is expected to end just as suddenly. I hoped that this blow would finish at sunset or at least during the night. The wind was blowing straight from Malta. We had the choice of heading either east or south. I chose to go south to keep away from the long, straight coastline of Sicily, hoping that before dawn we would be able to head south-east for Malta again.

The sirocco went on all day and all through the night. In the morning the barometer got a little higher, but so did the wind and the sea. We were not getting anywhere but I did not want to heave-to and start losing ground. We ran the engine to help the staysail and reefed main over the steep, high waves, and I wondered what I ought to do. I felt this weather had settled in for a long blow.

Before noon I had been able to get some good sights, and we heard Pantellaria on the direction finder. We were thirty miles east of Pantellaria, and south of the Terrible Bank, which was not a happy name. By motoring and sailing I thought that we could make good a course to Licata in Sicily, fifty miles to the east-north-east. According to the pilot book it was a good harbour for shelter.

We tacked and set off on our new course. It was no more comfortable than the old one, but at least we knew that we were going somewhere. I hoped that we would not arrive on the lee side of Licata and be forced to motor straight into the wind. The sirocco was blowing almost parallel to the coast and we would have no shelter until inside the harbour.

We crashed along all that day and into the night. The movement made life below uncomfortable, and there was a lot of spray on deck. I thought of Peter Vandersloot and his ability to produce good meals in bad weather. Our food on this trip to Malta was nothing like our Atlantic standard. It seemed to consist of a fairly steady diet of stale bread and cheese, tomatoes, garlic sausage and tinned meat. I had thought that leaving the catering to two Frenchmen would produce a cordon bleu cuisine, but that was not what happened. I resolved to go into things more carefully with the next cook.

Licata's light showed up where we had hoped to see it, and we were able to ease away and sail into the harbour. Michel was steering and I was with him as we neared the light. After a while I went to the saloon to try to catch an hour's sleep before we entered. Pierre was already sleeping there, rather precariously on the weather bunk. I thought of telling him either to go to the lee bunk in his cabin or to lash up the saloon bunk canvas to keep himself safe. But it seemed a pity to wake him, and there was not far to go. Already the sea was 'feeling' the land and the motion was less. I went to sleep on the lee side.

I was awakened by the sound of a tremendous crash. I swung off the bunk and jumped for the cockpit, my flashlight in my hand. I bumped into Michel, who was peering down into the saloon. He cried out:

'It's not up here. It's in the saloon. What has happened?'

I shone the flashlight around, first high up and then along the floor. We both burst into laughter. Pierre was lying there, as if in a strait-jacket. A sudden heave had thrown him out of his bunk. He had grabbed at the top of the swinging table, breaking the bolts which held the top to the framework. He hit the floor first and dragged the table-top onto himself. It was a folding top, with a centre and two leaves, and it closed itself over him. He lay like a warrior under his shield, unable to move his arms, his face and feet sticking out of each end while soup, glasses, and tomatoes rolled about his head. We rescued him and got ready to enter the harbour.

We anchored in the outer harbour, hung out the riding light, and went to sleep. In the morning a customs launch came out to us and cleared us without inspection. The officer told us that the sirocco would probably go on blowing for several days more, and he showed us where to go in the inner harbour. We moved there after breakfast, and lay stern-to below a tall quay.

We soon began to wish that we had stayed at anchor further away and put the dinghy in the water. The harbour was a safe and comfortable one, but it had a big drawback—sulphur. Along the quay, above us and to windward, went a continuous stream of carts. They were small, two-wheeled carts, each drawn by a pony with a tall, coloured plume of feathers on his head. The drivers stood like charioteers and drove like Ben Hur. This was all very picturesque, but the carts were loaded with loose sulphur for a steamer further down the quay, and the powder blew down-wind in a yellow fog, getting over and into everything.

As well as the sulphur in the carts, there were vast heaps of it in storage dumps at the east end of the port. From the carts and the dumps it spread itself everywhere as either a clogging dust, a suffocating smell or a stinging gas. Most of the people ashore and in the ships at the quay wore handkerchiefs over their faces and goggles over their eyes, but even this did not stop the tears and soreness.

If it had not been for the awful smell of sulphur and the feeling of the shortness of time, I would have enjoyed Licata. It was my first visit to an Italian town. Again, it followed the Mediterranean pattern of having the minor part on the sea and the real town a little way inland. I found that with some effort my bad Spanish could be turned into even worse Italian, but I was able to understand and be understood. Most of my information about Licata came from a barber who had set up his chair in the open, sulphurous air on the quay nearby and combined his hair-cutting with checking the loads and way-bills of the pony carts.

It seemed certain that the sirocco would go on for two or three more days. Michel and Pierre took this chance to go off for a night with the Palermo cousins, leaving by an afternoon train and coming back late the next evening. I sorted out things in the boat and tried to buy fuel at transit rates. I failed to do this, but had enough to wait until Malta.

Wandering around the port and the town on my own was amusing. Licata was not a holiday resort, and a visitor was a thing of interest to the inhabitants. People were helpful and kind to me, and my time was soon occupied. A very friendly electronic engineer took charge of me and showed me something of the town and the country. He drove me out to the old Castello Falconaro, and then back to spend the evening in his father's house at the top of the hill overlooking the harbour.

My crew came back full of enthusiasm over their Palermo visit. Pierre's cousins had been charming, their big house delightful, and their tales of Sicily and the Mafia intriguing. The boys were especially enthusiastic about the daughter of the house, MarieSol di Camporeale, and Susan Forster, an English friend staying with her. Michel tried to persuade me to take the two girls with us to Malta, and I think he had already told the girls that it would be all right.

I was against the plan. It looked as if the Malta trip would be an uncomfortable one, because even when the wind slackened there would probably be a big swell left for a couple of days. I already had one feverish and one seasick Frenchman and I did not like the idea of adding an unknown Sicilian princess and a fox-hunting English girl.

Michel at last accepted my refusal, but went on to say:

'I hope it is all right, but Pierre and I have asked the girls to drive over for lunch tomorrow. They want to show us some of the country and the Greek temples.'

I could see the thin edge of the wedge being applied, but it was too late to protest.

17. Netting the tunny—turmoil in the net

18. Netting the tunny—gaffing the fish

19. Cala Malfatana, Sardinia
20. Zola, Cephalonia, Ionian Islands

The sirocco was still blowing the next day, which was Friday. The two girls arrived very late for a lunch over which Michel and Pierre had made an unbelievable effort in the form of a very good Salade Niçoise. We had a happy afternoon, first on board and then driving to Agrigento over rolling country through almost treeless hills with fields of thin, light green crops. Summer had not yet brought its dryness and the fields and road verges were covered with flowers. There were many herds of goats grazing freely, which probably had something to do with the lack of trees. These goats were strange creatures with long straggly coats and enormous horns, spiralling straight upwards, bigger than I had ever seen. There were few villages, and each isolated farmhouse was like a small fort, with squared walls, no windows and only one approach into the central yard, through a massive gate.

I saw my first Greek temple before reaching Greece. It was a huge one, built about 400 B.C., on a hill looking down on to Agrigento and the sea. Its setting was splendid but the atmosphere was spoiled by the crowds of tourists coming and going in bright-coloured char-a-bancs.

By the evening the thin end had been driven deeper, and the wedge had broken down my resistance. Partly because our guests were attractive and amusing, and partly because I felt my crew had been having a miserable time and deserved a reward, I agreed to take the girls to Malta. Then the problem increased. In fact, it doubled itself. I learned that MarieSol's mother would not let the two girls come on the trip unless MarieSol's brother Paolo came as well. And Paolo's wife would not let him go unless she came with us. This was going to make things difficult, but there was such a mess already that the added complications would not make an impossible difference.

Then I laid down some conditions which I hoped they understood. The first was that I was going to sail on Monday even if the weather was bad, as long as it was not dangerous. But it might be too bad for passengers. They would have to come to Licata on the Sunday afternoon and stay in an hotel, ready to sail very early the next morning. In Malta they would have to stay ashore, to be out of the way while work was being done in the

6

boat. On the way back I would have to drop them at Augusta, on the far side of Sicily from Palermo, because I would have to wait there for my English crew to meet me. I thought that maybe when they got home and talked all this over they might decide the whole idea was too much trouble.

The wind went on blowing hard. It was still blowing on Sunday, but I felt that it was drawing to its end. So did some of the fishermen, and they told me to see what it would be like at midnight. If midnight brought a calm, there would be a gentle west wind the next day.

Sunday afternoon came with no sign of our passengers. An hour or two after dark we began to think that they must have become sensible rather than adventurous. Then a car stopped at our stern plank and they poured out, loaded with flagons of wine and boxes of food.

They were four people of completely different appearance. MarieSol was dark, small and Latin in looks with vivacious speech and gestures. Susan was quiet and very English with light brown hair. Paolo di Camporeale was surprising in not looking at all Italian. He was tall, broad-shouldered and very strong, with a slow, quiet voice, and curly, red-brown hair. His wife, Costanza, was just as surprising. She was very slim and delicate, seeming taller than she really was, and was fair-haired with blue eyes. She was the last to get out of the car, and as she stood at the end of the plank, I was sure that this ethereal creature was coming on board wearing high heels, a pearl necklace and a mink coat. Costanza has often since assured me that this was not so, and that she has no mink coat, but that was certainly the impression she gave.

Leiona was now more than full. We had seven people for only six berths, but I had an answer to the problem. I put Paolo in the fore-cabin with Michel and Pierre. I moved myself into the saloon and put all the girls into my small stern-cabin. This meant that each cabin had three people with only two berths, so we ran on the 'hot bunk' system. I made out a watch list, putting two on watch at a time, one from each cabin. The person coming off watch would get into the bunk vacated by the one who had just gone on, and everyone would have a place, in the cockpit or in a cabin. Paolo and Costanza were on watch together, Michel was with MarieSol, and Pierre with Susan. I had no watch and had the saloon to myself. This sounds a luxury, but I found I was needed on deck most of the time, as only Michel really knew what was happening.

At midnight I went to the breakwater to see what the wind was doing. There was not the calm that I had hoped for, but there was much less wind. It had increased a little by the morning, but it had lost the force of the last seven days. Pierre went off to warn Paolo, and we got ready to sail.

No fishing boats were going out, and local advice was that we should wait for another day or two. I felt sure that the sirocco was finishing and that we would be all right on our passage. At least, I felt sure that *Leiona* would be all right, and her crew, but I was not so happy about our passengers. Nor was poor Paolo. He had difficulty in getting their passports stamped for leaving, as it was early in the morning. When at last he found the Captain of the Port he was told:

'You must be mad to go out sailing in weather like this. All right, then, go if you must, but we aren't going to try to help you when you start drowning.'

Paolo was responsible for the safety of his household and felt that the sensible thing was not to go. I learned a year later that each one of them had thought the same way, but no one wanted to be the first to say so.

As it happened, we had a good passage to Malta, close-hauled all the way with full sail. There were only Force 3 and 4 winds, but the sirocco had left a disturbed sea which took some hours to settle down. The girls were sick at the start, but by the afternoon everyone was happy. We sailed under the great walls of Valetta Harbour and anchored off the Customs House in the morning.

We had lost five days at Licata so I wanted to sail the next day and there was a lot to do. I sent all the others ashore to do their sight-seeing and to leave me in peace to buy and load the stores. The agent had about £50 worth of tins and bottles waiting for me, and this amount exactly filled every bit of space in the lockers.

We dined ashore that night, and our passengers slept in an hotel. In the morning we filled the fuel tanks and sailed. I was sorry to leave Malta so soon, with its impressive harbour and interesting history, but the days we had lost through bad weather at Mallorca and at Sicily left us no time to spare.

From Malta to Sicily we had 'bikini weather' all the way. It was good for swimming and sun-bathing, but not for sailing. We had to use the engine for twelve of the twenty-two hours.

Costanza finished her watch at midnight. She asked me to call her early if I could show her a beautiful dawn. And it was beautiful. While we were still in a dark sea, the sky in the east was growing light and a pink glow was spreading along it. High up and ahead the early rays of the sun were lighting up the small clouds and shining on the peak of Mount Etna, standing in clear air above the clouds, capped with gleaming white snow, and trailing a plume of white steam. I woke Costanza.

As the daylight grew, the others began to wake and soon were all on deck or in the cockpit. Paolo made coffee and that led to breakfast. Costanza leaned down to look at the saloon clock and said to me:

'Keith, the clock has stopped.'

I looked and told her it was going. She sat up.

'What? It is really working? But I have been up and looking at the volcano and talking and having coffee and breakfast and it is not yet seven o'clock! No, I do not believe it. This has never happened before. It is not possible.'

We reached Augusta before noon, with enough wind to take us in under sail. Near the Port Captain's office was a space to go alongside, and we made for it. To my surprise Richard and Diana were standing there, two days before I expected to see them. They had been just as surprised to see *Leiona* sailing in.

They came on board, and were soon followed by port officials and police. For an hour the boat was an overcrowded turmoil while entry formalities, packings and unpackings, greetings and farewells, went on in three languages.

Then Paolo bundled all his household, including Michel and Pierre, into an outsize taxi, and they set off to make their long way to Palermo.

We had arranged to meet him and Costanza somewhere on the northern coast in June to come sailing in the Lipari Islands, of which Stromboli is one.

In what now seemed a quiet and deserted boat the three of us sat down to work out our plans for the summer, starting with the charts of the Strait of Messina.

CHAPTER THIRTEEN

Volcanoes and Whirlpools

We spread the charts over the saloon table and planned the start of our route. Our first three landmarks were the three big volcanoes of the Mediterranean. The first one, Etna, had brought me to Sicily from Malta, but we were to sail past it for yet one more day. Then there was Stromboli, in the Lipari Islands, and then Vesuvius, above the Gulf of Naples, where the half moon of the bay had Ischia and Capri at either tip.

My crew had already seen the east coast of Sicily during the last week, so from Augusta we went north, passing Taormina and leaving Etna behind us, and entered the Strait of Messina.

The names of Scylla and Charybdis appear on the chart. We found in the *Odyssey* the descriptions and warnings that Circe had given Ulysses about the six-headed monster on one side and the engulfing whirlpool on the other. Other warnings were given to us by the more matter of fact *Admiralty Pilot*, but even those gave us something to think about.

In these days most people pass through Messina in large ships which are not troubled by its strange currents and eddies. Our boat was about the size of the one Ulysses sailed, and not much faster. We had to take notice of the four-knot tidal streams and of the steep seas they could form against the wind. Warning was given of the three largest whirlpools, shown on the chart, which could be dangerous to small craft. It was comforting to read that Scylla was now very feeble, and that even Charybdis had become much less impressive than in former times. But to stop us taking the Strait too light-heartedly the *Pilot* added:

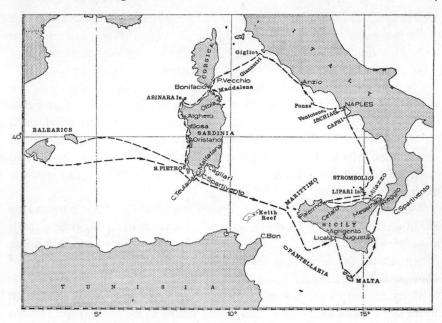

Fig 7 The Italian Sea

'. . . moreover, in the vicinity of the high land, on either side, vessels are exposed to violent squalls which descend through the valleys with such strength as, at times, to inconvenience even steamers.'

We wanted to go through the famous Strait by daylight, both to have a good look at the whirlpools and scenery, and because all that I had read made me frightened of trying a night passage. The *Pilot* and *Brown's Almanac* showed me how to find the favourable tides, but for my first time through I wanted to ask local advice.

Reggio was a convenient place to get this advice, so we stopped there for the night. I had been told that it was a better and safer harbour than Messina, but we found Reggio bad enough. At first we went alongside some wooden piles, but soon changed to lying stern-to. A heavy surge was raised by the wash of each fast hydrofoil ferry as it came in from

Messina. While the craft was moving fast on its small floats and tall legs it did not make much disturbance, but as it slowed down, lowering the large hull quickly into the water, lines of waves shot across the oily surface of the harbour and struck against the piles.

The Harbour Master and a police official came on board at once. They were so pleased at finding someone as pretty as Diana in the saloon that they settled down for a long visit. The ship's business was soon completed, in their bad English and my worse Italian, and then the official visit turned into a party. It was late by the time they went, but we had managed to find out a lot about Reggio and the Strait, and to get our chart marked with the best route.

We had been told to start at 7 o'clock, to be clear of the narrows by 10. In the early morning a north-west wind was blowing at Force 4, with a lot of white water in the Strait. It didn't look right to me, but some fishermen on the quay said that it would be safe for the passage and confirmed what we had been told about the route. We followed the marking on our chart, motoring all the way, keeping close to the Italian shore until Punta Pezzo and then going north to close with the Sicilian side. We could see the long sweep of the overhead cable that crossed the narrows at Cape Peloro. The wind died away as we neared Pezzo, and the water became flat.

As we came to the centre of the narrows a strange boat crossed our bows, heading for the Sicilian shore. It was not very big and it had a tall, thin scaffolding amidships, like a tower. It looked like an exaggerated imitation of the raised mobile platforms used for working on tramline cables, and for one stupid moment I thought it was something to do with the overhead cable ahead of us. It obviously was not that, which would have needed a platform more than two hundred feet high, but we could not make out what this queer boat was.

The mystery was solved when we came near to the far side. Lying at anchor close to the beach were a dozen of these boats, some about forty feet long and others slightly smaller. I had heard of the swordfish boats of the Strait of Messina, but I had not imagined them to be of such a strange design. They were long, low, flat-decked boats with enormous masts and bowsprits. In the smaller boats these were normal wooden spars with extra rigging, but the larger boats had metal structures like box-girders.

Large or small, the proportions were fantastic. A forty-foot boat had a bowsprit about sixty feet long, and a mast of seventy feet, with a complexity of stays and shrouds. At the end of the bowsprit was a pulpit where a man stood with two harpoons, while in a crow's nest on top of the mast stood another man, able to see deep into the water from that height and put the harpooneer over the swordfish. This masthead man was not just a lookout, as we had thought at first, but he steered the boat with a small wheel and a long steering cable like a bicycle chain. One boat had three men at the masthead.

These boats were very fast, and darted about as the elevated helmsman thought he saw a target. We cruised around for a while, trying not to get in the way of a chase, but we did not see a harpoon thrown. Later on, in Milazzo, we had swordfish to eat. It was very good. Having seen the amount of effort and lack of success, we were not surprised that it was also very expensive.

The passage through the Strait was almost disappointing in its lack of difficulties or dangers. We wandered around looking for whirlpools, and at Charybdis found two mild eddies which tugged at *Leiona* and turned her from her course. I was glad that we had been able to choose our time and find a calm day. I would not like to have met even half the discomforts which the *Pilot* describes.

Clear of the Strait we turned west and sailed along the north coast of Sicily as far as Palermo. All the way we had a light wind against us, and I was tempted to make a proper use of it and go north into the Lipari Islands. But we were in no hurry, the weather was good, and the sea was warm, so we tacked slowly westwards. We had two good reasons for doing this, to have a good look at the Sicilian coast, and to meet Paolo and Costanza and go sailing in the Lipari Islands with them. Palermo seemed the best place for this meeting.

On our way from the Strait of Messina we only called in at one place, Milazzo. This north coast of Sicily was something like the north-west coast of Mallorca, though not so sheer. We sailed for four days past the background of an unbroken mountain range, along a coast without any safe anchorages or harbours. This was slow progress for covering a hundred miles of coast, but our distance through the water was nearly double this, as we were tacking all the way. It was slow but enjoyable.

We fell into a delightful 'lazy programme' of sailing. Close to the shore we would tack and then sail out to sea for three or four hours, to tack again and sail shoreward. At night there was hardly any wind, but just enough to keep us steady and on our course, even if we were not getting far. With the dawn the breeze would strengthen, and we would sail pleasantly until noon, when the wind almost died away. This was the time for Diana to lay out a cold salad lunch, and for Richard to get ready a tray of iced drinks. Then we would swim, have lunch, and all go to sleep, leaving *Leiona* to sail herself. In the early afternoon the wind would come back again and the movement of the boat would wake us. Then we took sailing seriously and made some progress until sunset brought again the calm of the night.

We saw a few steamers each day and night, usually well to seaward of us, but we only saw one yacht. One afternoon, when we were all below, we were startled by the sudden noise of what sounded like a telephone bell. Fifty yards from us and going the opposite way was a blue German ketch with two men on board and the wind behind her. They laughed when they saw us all appear, and called out:

'Good. We are happy to see you are all right. For nearly an hour we have been looking at you. We saw the boat sailing straight, but no man anywhere. We thought maybe everyone had fallen overboard.'

We waved at each other and they ran on their way to Messina while we went on tacking for Palermo.

We tried to time our tacks to suit the cook. The galley and sink were to port, and so was the refrigerator. This made the starboard tack the 'cook's tack', as it was easier to work in the galley with the boat heeled to port, and the refrigerator could be opened without danger of bottles and dishes sliding off its shelves.

Milazzo was a more pleasant place than the official description had led us to expect.

The only ugly thing about the harbour was an oil refinery two miles along the coast, but it was not necessary to look in that direction very often by day, and at night the lights on the chimneys and the glare of the flames had a certain beauty.

The buildings around the harbour were modern and the new town was simple but without any special charm. The harbour was sheltered from the west by a long, tall peninsula running northward. The old town was built on this peninsula and a walk along it was full of interest. We walked along the wide boulevard beside the sea and then climbed up through the narrow streets and old houses. Higher still, we came to a cemetery where the family vaults were larger than many of the houses we had passed. Highest of all was the big citadel from where we could look both ways along the coast and out to the Lipari Islands. Vulcano, the nearest, was only fifteen miles away.

In the old town Diana found places and people to paint, and we stayed in Milazzo five days while she filled her canvasses. We enjoyed strolling about the town and Richard was delighted with the good and cheap little restaurants we discovered. He was really a very conservative person, especially over his food, but he was gradually being broken into Mediterranean habits.

It was in Milazzo one afternoon that Diana started to laugh, saying: 'Keith, I don't know what you and *Leiona* are doing to my Richard. Lifetime habits are changing. This morning he was happy with new bread instead of toast for breakfast, and look at him now. He is eating figs straight out of the barrow, unwashed and unsterilized.'

He still held out against having broken lumps of ice put in his drinks. Watching how the blocks were being tipped out of a cart and dragged across the road in front of our restaurant that evening made me think he was right.

On our slow passage our inshore tacks often brought us to beautiful places. The big monastery on the mountain above the Gulf of Patti was a wonderful sight, at the mouth of a valley of farmland, copses and vineyards. And one afternoon we tacked under the Norman cathedral and walled town of Cefalu, where the walls climbed upwards and outwards from the shore to join vertical cliffs

Finally we rounded the huge headland of Zaffarano and motored into Palermo, to arrive there as the sun set. In the big harbour we shared a steamer dock with the beautiful barquentine *Palinuro*, an Italian sail training vessel. She was a striking contrast to the modern buildings in the background and was a fine picture with her black and white hull, buff spars and golden figurehead.

Even in June it was hot where we lay below the level of the dockside, shut off from the wind. This was not much of a worry, because we spent very little time in the boat during our week in Palermo. Paolo and Costanza adopted us and life became a busy social whirl. We were swept from house to house, from luncheons to cocktail parties, and from cocktail parties to dinners. It was fun, but it was exhausting and bewildering.

The best days were when we were taken into the country, over the mountains to the wide plains which sloped away on the other side. There we saw the farm estates, first of Costanza's charming parents, and then Paolo's own farm. Each big farmhouse was built in the same Sicilian style, like a fort, with only one entrance to the central courtyard, and no ground floor doors or windows on the outside. The family lived along the front wall, on the upper floor, while the servants had the back wall. The lower floor and the ends of the rectangle were filled with stables and barns and storerooms.

Our week was full of interest and amusement, but good sailing weather and time was going by, and I was glad when we could start for the islands. It was good to see Paolo and Costanza coming on board to sail again, this time as friends and not as strangers.

Paolo's arms were full of harpoon guns, goggles, and flippers, and he carried an enormous flagon of wine from his own vineyards. Costanza no longer gave the mink coat impression. The whole of her wardrobe for a week's sailing was the shirt, trousers and sandals she was wearing, a toothbrush, a comb, a bottle of sunburn oil, and four bikinis wrapped in a towel.

We had some good days in the Lipari Islands. The only trouble was finding places sheltered enough and shallow enough for safe anchoring. All these islands were volcanic peaks and their sides fell sheer into deep water. The one harbour was at the town of Lipari, and it was too busy and crowded for our liking.

We spent one evening in Lipari and wandered over the old town. Within the walls near the citadel was a big church and a monastery with beautiful gardens and thick hedges of rosemary and hibiscus. It was the place where I first knowingly saw how capers grew. I had only thought of them as rather unpleasant little black berries in tasteless sauce for boiled mutton, but in these islands the caper bushes were the most common covering for the steep hillsides and the only bush that stopped the higher slopes from being barren black and dark brown earth.

The moon was full when we moved from Lipari after sunset to the island just south of it, Vulcano. There we found two small bays at its nearest point, facing east and west, divided by a narrow headland. We anchored in the western bay in two fathoms over a bottom of flat black lava sand, and made it our base.

Until our last day we had light winds and calms, which were what we wanted for swimming and diving. To swim deep down with a mask along the vertical sides of the rocks with their waving weeds and coloured fishes was like moving past the windows of a lighted aquarium. In the few places where there was white sand, we could see down to the anchor lying between black rocks, but more often the chain disappeared into darkness.

The most impressive memory was of our last night in the Lipari Islands. We were then at Panarea. It was a clear, quiet night with a big yellow moon spreading its path of gold on a silver sea. We sat until late on the veranda of a little hotel at the edge of a cliff, with *Leiona* anchored in the moon-path, off the open beach. Twelve miles to northward was the black mass of Stromboli. The volcano was stirring that night and breaking the peace of the sky with sudden flares of bright red fire.

Stromboli's peak was in thick white cloud next morning. We sailed to the volcano in a north-west wind which made me think about reefing but never got quite strong enough. In front of the white houses of the village we put Paolo and Costanza ashore on the beach of black sand and set a course for Naples and the north, towards Vesuvius.

We had one night at sea and reached Naples after sunset on the second day. It had been an odd night. After leaving Stromboli we sailed well all day but at dark the wind dropped a little and strange little isolated lightning squalls started to appear all around us, with a six-foot swell from the north-west. None of the squalls passed over us and by dawn the sky was clear. The swell died down, but so did the wind, and we were left with a dull day of motoring.

To make up for the dull day, we had a striking entry to the Gulf of Naples in the evening. In the dusk Capri showed as a rocky silhouette while a hazy red sunset lit up the higher slopes of Vesuvius and the enormous clouds that were towering and spreading above the volcano.

We kept away from the harbour and anchored in the open bay near the yacht harbour of Sannazzaro, at the western end of the city. The moon rose red that night and we lay rolling gently in front of the city lights while cars drove along the seafront, two hundred yards from us.

Two days later, we left Naples. We were glad to get away from a big city. Even in the yacht harbour the water was thick with oil. To keep the dinghy clean we had moved from our first night's anchorage and gone stern-to at the quay among the other yachts. We discovered too late that because of various underwater rocks, it was impossible to be close enough to the quay to use the stern plank, so we had to use the dinghy as well, getting both dinghy and warps filthy.

A fast sail from Naples in the afternoon had brought us to Ischia at sunset. It was a wonderful sight with the magnificent castle silhouetted against the red sky. The next striking sight was a switch to the ridiculous. Above the small cape at the narrow harbour entrance a peculiar golden tower was showing itself. We could not think what it was until we entered the harbour and saw it was the stern castle of Cleopatra's barge, left over from the making of the much publicized film.

The harbour was built in the crater of an old volcano where the sides had collapsed. It was full of yachts. We motored in very slowly in the falling dark and were lucky to find an empty place. We sat for a while looking at the bright lights and the happy people strolling around the harbour.

Just then a tall figure stopped at the end of the stern plank and a surprised and inquiring voice said, '*Leiona?* From Antigua?'

It was Philippe Peissel, Michel's brother, whom I had met for a few minutes in Bermuda. He had come here to sail an English cutter back to England, and was our next-door neighbour.

Lying in Porto d'Ischia was a complete change from anything we had done so far, and we found it lots of fun for a while. People, Italians or others, were all friendly and easy to meet and we soon found that we had made many acquaintances.

It was all amusing, but in a few days we had had enough of it. There were the usual snags of harbour life. The windless days and hot nights were uncomfortable and we missed our swims from the boat. To walk half a mile to swim off a beach was not the same thing. Meeting friendly people was all right, but we were missing the peace of the small islands, and Ischia was noisy until far into the night. Then early in the morning the ferry steamers from Naples began to come in, and each one sent a series of waves surging around the quay wall, making the warps jerk and creak and the stern planks rattle.

We left, and sailed off to visit the small bays of Procida Island and in the south of Ischia. The Ischian village of San Angelo was very pretty and we stayed for three days in the wide bay beside it. On our first evening there Capri seemed almost beside us, although it was fifteen miles away. It looked beautiful, standing up in a clear outline above a base of mauve haze.

From there we sailed west to Ventotene and Ponza, staying only a night in each. It was a time of good settled weather, so we anchored off beaches instead of going into the harbours.

We passed close to Botte Rock on the way to Ponza. It looked like a square-rigged ship, like Sail Rock off the Grenadines, and we wondered if anyone had fought it with guns.

From Ponza we went north to Anzio. We closed with the land southeast of Anzio and were looking for a landmark on the long sandy beach, not easy to see in a hazy afternoon. Through the haze we began to see some buildings or tall, pointed rocks but could not make out what they were. They began to take shape as we got nearer, and we were amazed to see a large sphinx and some pyramids, more left-overs of the Cleopatra filming.

I had a purpose in going to Anzio, which was not an attractive port or town. Signora Maria Pera drove down from Rome to have lunch with us. Not many English people know her under that name, but as Mary Blewitt she must be admired and thanked by nearly every Englishman who sails a boat out of sight of land. Her book, *Celestial Navigation* is as much a yachtsman's bible as Eric Hiscock's *Cruising Under Sail*.

After Anzio we had a hundred miles of slow sailing with nothing of interest except two steamers uncomfortably close to us at night, and the muddy water of the Tiber. We were having our 'lazy programme' swim at noon, and I went underneath with a mask to have a look at the boat's bottom. I could see nothing at all. I thought the glass of the mask must be dirty, and came up to clean it. This made no difference. For the first time in the Mediterranean I was swimming in muddy water. It could only have come from the Tiber whose mouth was out of sight in the haze, some ten miles away.

We had not quite finished with volcanoes. Our last lonely anchorage was inside the crater of another dead volcano, the solitary little island of Giannutri, near Porto Ercole. We slept in a peaceful deserted cove at one corner of the crater. The air was clean and scented, and in the morning we were wakened by the singing of birds and the whistles of quails.

Elba should have been our last stopping place along the Italian coast, but we did not reach it. We were almost there when the wind turned against us, blowing hard from the north. By then we were in Giglio, where we had gone after Giannutri. It was a pretty harbour, with two villages in sight, one around the harbour and one high on the hillside above, bunched beneath the walls of an old castle.

The north wind went on blowing, and sent a swell into the crowded harbour. We were not comfortable and wanted to leave. It would have been even more uncomfortable sailing north to Elba, so we decided to leave it out, and we set off to the south-west, for Corsica and Sardinia.

CHAPTER FOURTEEN

To Corsica and Sardinia

Corsica made two sudden and surprising appearances. As it was only eighty miles from Giglio, and with mountains nearly 9,000 feet high, we thought that we would have it in sight for most of our passage. The air seemed clear, but we saw no mountains that morning or afternoon. Then at sunset a tall line of jagged peaks stood out black against the pink sky, and slowly faded into the grey of night.

The next warning of Corsica was to our noses and not our eyes. After midnight I was steering in what had become almost a flat calm, and I was nearly asleep. A strong sweet scent of flowers and trees roused me, brought in by a new wind from dead ahead.

Soon after that I saw a light. It was so bright and its flash was so long that I thought it must be a flare or the working lights of a nearby fishing boat. A check of its timings showed it to be the big light of Chiappa Point, at Porto Vecchio.

The eastern sky was growing light as we neared the mouth of Porto Vecchio's long entrance. Daylight spread as we passed between the flashing lights on the black rocks and islands of the difficult channel. Then we were inside a beautiful enclosed bay like a mountain lake, where pine trees came down to coves of white sand, and the surrounding hills made a background of green slopes in the south and of steep grey peaks to the north.

At the south end of the bay there was a small open harbour and a few houses, with a close-grouped grey town standing separate on a hill above.

At a low quay were two yachts and some fishing boats. We anchored near them and hauled our stern in to the quay. Then we got hold of a taxi to take us up to the town, talking French as a welcome change from our halting Italian.

Leaving a boat in a strange place, either swinging to an anchor or lying at a quay, always makes me feel worried, but this time the day was so calm that I felt *Leiona* was safe. We wandered around the town, and were having a carefree aperitif in a sunny square when I caught a glimpse of the bay through a break in the houses. The flat blue water we had left was becoming flecked with white. A north wind was starting to blow, straight into the quay at the open harbour.

We left our drinks, woke the taxi driver, gathered our baskets and parcels, and got back to the harbour as quickly as we could. The wind was not yet strong, but short little waves were breaking against the quay, and the boats were moving and bumping against each other. *Leiona* had some blue paint and an open scrape on one white topside, so we thought it was time to go. We motored across to the other end of the big bay and found a quiet cove for the rest of the day.

Our visit to Corsica was a brief one. I wanted to go further south as soon as we could. For no good reason, the latitude line of 40° north has a magic meaning for me, and I can almost imagine it as a visible and material barrier. I know this is ridiculous, like my idea of sailing 'at right angles to the world', when I have the comforting sensation of moving safely along lines which exist on the chart and seem to exist on the sea as well. Even so, I have a private conviction that once I go north of the 40° latitude I start running into trouble and bad weather. I cannot rid myself of the feeling that I am better off south of this line, and in any case it includes most of the areas in which I want to sail.

In the Mediterranean it includes all the Balearics, the southern half of Sardinia, Sicily, Corfu, and nearly all the Greek islands. Further afield, in the Atlantic it includes the West Indies and Bahamas, Bermuda, and even the Azores. I shall have to overcome this prejudice somehow when I want to go to Istanbul and into the Black Sea or up the Adriatic.

My silly conviction was driving me to go south, but I wanted first to go through the Strait of Bonifacio. We let the wind choose for us again, to see whether to go through the Strait as our next move, or to go to the north-east corner of Sardinia. We were attracted by the complex of deep bays and islands in this part and we wanted to spend a month there.

The wind made a firm choice. The next morning we saw the clouds moving westward. I had been told that the winds in Bonifacio were nearly always from the west, so this gift of a following wind to take us through was too good to let go. So we sailed for the Strait and the town of Bonifacio.

The wind held all day, giving us a quick and comfortable sail, and a beautiful one. Against a background of green and grey mountains the islands and cliffs were pink and shining white above a deep blue sea.

Bonifacio's entrance was a high narrow cut, hard to see where it ran straight into a cliff face and then turned sharply to run between an alley of tall white cliffs, with the small town and a low quay at the end. The whole was dominated by the square-cut citadel of the old Roman fort, which a Foreign Legion garrison held.

We stayed six days in Bonifacio, kept there by a strong west wind which started the day after we arrived. We found plenty to do and we enjoyed our stay, trying out different restaurants and bars, climbing to the top of the narrow ridge between the harbour and sea, or just looking at boats. We saw the house where Napoleon had lived as a boy, and we talked with the people of the town and of the other yachts, one each American, French, German, Spanish, and English. And I got a mechanic to have a look at the engine's self-starter. He brought a helper with him, a Czechoslovak in the white kepi of the Legion.

Diana tried to do some painting, but it was not easy. The boats were safe enough in the harbour, but the wind sent gusts and eddies down the alleys and around the corners, with whirlwinds of dust. The efforts of re-siting easels and retrieving flying canvasses became too much, and Diana gave up her painting and joined Richard and me in sight-seeing.

When we sailed we went back through the Strait again, this time with a west wind to help us. We went first to La Maddalena and then sailed in the islands and bays of the indented coast between there and Olbia. It should have been fun, but we had bad luck. For the first time we had a long stretch of bad weather, not so much of strong winds as of wet ones, with low cloud and rain.

It was not what we expected in a Mediterranean July, and we began to feel depressed under the black hills and grey clouds. We decided to change coasts and to move south. We went back to La Maddalena for a day, and then our luck and the wind changed. The wind turned east and the sun came out. We sailed on to stay for one night in a deep-reaching quiet harbour at the most northern point of Sardinia, Santa Teresa. From there we turned south and worked our way down the west coast.

The Forty Degree Magic began to take effect even before we got down to that latitude. As soon as we started going south the grey skies, rain, and cold wind left us, and so did our feelings of depression. Even *Leiona* moved more lightly and happily.

We coasted down the wide Gulf of Asinara, which forms the north-west coast of Sardinia, anchoring once off an open beach, once in the tiny harbour of Castel Sardo, and once in the curve of Asinara Island.

Asinara Island is long and thin, at the north-west tip of Sardinia, separated from the mainland by a narrow channel ten feet deep. The channel is not an easy one, twisting through reefs and sunken rocks, not marked by buoys, but by two pairs of stone pillars as leading marks. Going through would save the big sweep north of Asinara Island, and looked interesting. We had a calm bright day and clear water, so we tried the channel and motored through.

It was an interesting passage. I found it frightening, but acceptable in fine weather. The first pair of marks led us into the narrowing channel, flanked closely by shallows and hidden reefs. The second pair of pillars marked a sharp change of course, and we had to 'turn left' like a drill movement. These marks copied the Duke of Plaza Toro and 'led' us from astern, which I found a hard system. By keeping them in line, helped by the compass, we came to the open sea, after passing uncomfortably near a reef over which a slow but large swell was breaking.

Our next town was the pretty port of Alghero, which Diana found a good one for her paints. We saw its towers and walls after rounding the magnificent headland of Cape Caccia. New hotels were being built to the south, but the old town and the harbour still looked medieval.

The entrance was very narrow, a long channel leading between a rock and a reef to port with the fortress wall close above our starboard rigging. We lay at a quay among the fishing boats, in front of the bastions and gateway of the thick walls.

Alghero was a pretty harbour, but not a safe one, if a west wind blew strongly. We even found our anchor dragging when the wind was from the east, off the town, and I did not like leaving *Leiona* there for long. In spite of this, it was a good base, because safe and comfortable shelter could be found in the pleasant bays near to the town.

We had a special liking for Alghero because of an act of kindness to us when we first arrived. We reached it on a Saturday evening, when the only money we had was in travellers' cheques. These are not always easy to change in a strange and small place, but we set off to see what we could do. Maybe if we had taken the long walk to the new hotels at the far end of the town we might have cashed the cheques. In the old town near the harbour we met with no success.

Then in a square inside the big gateway I saw a taverna with the sign 'El Rey Carlos Primero'. The Spanish title caught my eye, and while I was looking at it the owner greeted me with a 'Buona sera'. I asked him for advice on cashing the cheques, but he said only the banks could do that, on Monday morning. I pointed out that we wanted money to pay for dinner that night, and to buy food for the next day. A young fisherman stood up from a table where he was sitting with two friends, and began to feel in his pockets. He started laying money on the table, and so did the other two. Then he came over to us with a handful of money, saying:

'I hope this will be of use to you. I think you are from the white

English yacht. My name is Giovanni and my boat is near yours, the *Santa Maria*. You can pay me back later, when you have the money.'

We were grateful for his help, and trust. Now that we had some money, even though it was his money, we could ask him and his friends, who were his crew, to drink with us. We learned from them, with some difficulty, that Spanish names and some knowledge of Spanish were not uncommon. Alghero had once been captured and colonized by the Spanish, and the people were proud of their old connection with Spain.

We were sorry to leave Alghero, but we had dates to meet and moves to do. Richard wanted to be in Rome before September, and it was now nearly August. I had not found a new place for spending the winter.

During my wanderings this summer the problem of my winter quarters had been on my mind, but not as urgently as in the year before. So far I had not seen anywhere that I liked as much as Puerto de Andraitx. One little fishing boat harbour near Palermo, Porticello, had been possible, but it looked too cramped and crowded. Malta was a practical choice, but I felt that the harbour was too big and the country too barren. I liked Alghero, but I was not happy about its safety.

There were still a few more Sardinian ports to see, but I had already begun to think that my best course would be to return to Andraitx. I knew it and liked it and felt safe there. Anyhow, the distance between Sardinia and Mallorca was not great, only three or four days of sailing, so for next year's programme it did not make much difference in which island I stayed for this winter. I could do my preparations more easily in a place and country I knew than by starting afresh in Sardinia or Sicily.

I decided to go back to Andraitx, but there was still plenty of time ahead. We talked over our next moves. Richard and Diana could come with me to Mallorca, and then fly to Rome from there. Or we could all stay in Sardinia until it was time for them to go, and then I could go off alone to the Balearics. This seemed much the best plan, giving us all another month in Sardinia, where we still had lots to see, and leaving them with only a short journey to Rome.

At first they were unwilling to let me sail off on my own, but I managed to persuade them that it was a short and easy open sea crossing, and that I liked sailing alone. I had not had the chance of sailing single-handed since I first sailed *Leiona* in the West Indies, and I was looking forward to it.

We had one more month of pleasant and leisurely sailing after leaving Alghero, all in the west and south of Sardinia. We liked it better than the more fashionable north-east. The country was more gentle and green, although the hot summer had already done much to making it dry and brown. The long sweeping hills and valleys of the centre and the south looked more friendly than the jagged rocky peaks of the north.

Our 'lazy programme' fitted well into our move south from Alghero, and we found an anchorage at the end of each day's sailing. We shared the round cove of Bosa Marina with a lovely trading schooner, and sailed with her in the morning, on our way to the big double-headed Gulf of Oristano.

Many of the anchorages were too pleasing to leave quickly, and we had time to stay. Our longest stay was in the very south of the Gulf of Oristano. Here was the Stagno di Marceddi, a thin, shallow inlet four miles long. It was seldom more than a fathom deep, and often less than a metre. The narrow entrance channel was marked by poles and bushes, and our wake rippled over sandbanks on either side.

There was a modern village north of the entrance, and an old one south of the inlet. Teams of fishermen were always busy netting over the shallows. Each team had four fast boats, moved by outboard engine, oars, or poles, and they used them to lay out an intricate, symmetrical, and enveloping pattern of nets, set up on stakes. They worked very fast, two boats on each side separating to meet the boats of the opposite pair and close the net. As soon as they had laid the net they swooped round to gather it in, always with a good catch, and then moved away to plant the stakes in a new place.

The last town we visited in Sardinia was Carloforte. This might have done for a winter harbour, but by then I was set on going back to Andraitx. Carloforte was on San Pietro Island, at the south-west corner of Sardinia. Oddly enough it too had been colonized by foreigners, this time by the inhabitants of the Tunisian island of Tabarca. They had been brought to Carloforte by the Italians for safety from the ravages of pirates.

The town was pleasant, and so were the people. We anchored in the southern part of the harbour, which we found quieter and safer than the quay. Scottolino, a boatman, was a great help, and Nicolo, the enthusiastic young barman of the only hotel, was a useful informant and guide.

Nicolo told us that there was to be a '*matanza di tuna*'—a 'killing' or 'massacre' of tunny fish—and urged us to go to see it. He came with us in *Leiona*. It was a calm morning and we motored north to where the nets had been set to turn the travelling tunny shoals towards the shore. The first nets were not for catching the fish but only to guide them to the right place. I had read in the *Pilot* books of these nets off many Mediterranean coasts, where they could be a danger, reaching several miles out to sea. We had met one lot by day near Gibraltar, on my first entry to the Mediterranean, and had seen the danger. The nets were big ones, hung from a headrope three or four inches thick, which was floated by huge lumps of cork, blocks of wood and some old heavy boats. The seaward end of the nets was marked by a flag and lights, but these were not manned and could disappear in bad weather.

By the time we reached the nets the fish had been enclosed in the 'chamber of death'. This was a net with a bottom to it, unlike the turning nets, hung from heavy black open barges lashed together over an area about the size of four tennis courts. The entrance to the chamber of death had been closed by a weighted net, and eighty or ninety men standing in the barges at two opposite ends were hauling the net, raising its lower part, now full of fish. This raising was a slow business as the net was heavy and deep.

In the middle of the enclosure was a light boat with two oarsmen and a tall and distinguished-looking old man who directed the hauling. His directions and his gestures were like those of the conductor of an orchestra. The whole process was dramatic and operatic, even to the chants of the haulers who were sometimes stirred to a frenzy by the gestures of the conductor.

In the small boat and around the barges there were piles of oilskins. The reason for these was soon clear. Excitement grew as the bottom of the net came near the surface. The water outside the barges and inside the net was clear and calm. Suddenly a dark shadow darted for an instant within the net. Then another. Then several, their tails and fins cutting the surface. The fish were big, from four to eight feet long. Soon the whole of the surface inside the net was a flurry of white foam and spray while the fish fought to get free. Blood was mixed with the foam from the wounds of fish which had torn themselves against the net.

The next stage was the gaffing of the tunny. The hauling stopped and each man took up a long pole armed with a wide gaff. One man would reach out and strike his gaff into a fish and pull it to the barge. When it came within reach, two, three or four men would dig in their gaffs and they would heave together, dragging the fish over the side of the barge and letting it fall into the hold. The gaffing was not a pleasant sight, and we left before it was finished. Nicolo said they hoped to land eight hundred fish or more.

San Pietro Island had many pretty coves to visit for day sailing, and the waters of the wide channel were very clear over shallows of white sand and ravines of dark weed. For our last week of sailing together we went as far as Cape Spartivento, at the south of Sardinia. We found good anchorages all the way, but the most beautiful was the last bay, Cala Malfatana, near Spartivento.

The hills around the bay were like the ridges of Mallorca's Cabrera, and we walked all over them. We tucked *Leiona* in behind a small headland, off the sandy beach of a valley's mouth, near a village of five houses.

We shared the bay with a party of wandering fishermen who became our friends. They had two large and two small open boats, and anchored each evening close under the cliff of the headland, cooking their meals ashore on the strip of pebble beach.

They were fishing in the bay, and their catch was taken away each noon by a man with an old truck. They fished by stretching a long net right across the bay between the two big boats and then moving in towards the beach, closing in as they neared it. At the end they would wade in to haul the purse of the net into the shallows and up on to the beach. We used to help them land their catches, which were very small.

Then it was time to go back to Carloforte and finish our Sardinian sailing. I watched my crew go off in an early morning ferry, and I sailed that same day for Menorca.

The passage was without troubles. I saw no ships by day or by night. Sometimes a pair of shearwaters would circle around me, and sometimes there were dolphins to look at, but otherwise I had an empty sea the whole time except for one afternoon.

I used to stay on deck all night and do my sleeping in short stretches during the day, leaving *Leiona* to sail herself, and coming up every hour or two for a look around. In one of these looks I saw something flashing in the water a mile ahead. I thought it must be a barrel or large piece of wood. There were some more flashes, and then first one and then a series of explosions of white steam. As I got closer I saw that the flashes were from the shiny wet skins of a pair of whales playing on the surface. Their spouts were like small round clouds spreading forward at an angle from the end of their blunt, squared-off noses. I could see long, thin lower jaws opening, and twice a tail was raised, with wide pointed flukes branching from a slender root. I think they must have been sperm whales, but I had not heard of them in the Mediterranean.

After three days I saw the grey land of Menorca low ahead. There had been one day of just the right amount of wind, when *Leiona* steered herself unaided, and one day when the wind was too light. And one day when the wind had been too strong for comfort, but it had come from astern and helped me on my way.

I sailed into Port Mahon. My summer of sailing in the Italian sea was over. I moved on slowly down the Mallorcan coast and back to Puerto de Andraitx.

Here I spent another happy winter, but in a different way. This time I moved *Leiona* and myself away from the land, and lived on board, lying to two anchors in the middle of the harbour.

The same friends were there, and the same things to be seen, but more clearly with the help of knowledge.

As before, the winter seemed too short to do all the things I wanted to complete, and to prepare for moving my boat for some years in the waters of Greece.

CHAPTER FIFTEEN

Ionia and Corfu

The Balearic spring came with its colours and perfumes of flowers and blossom and the fresh green of new leaves. In April I was all ready to sail with *Leiona* again, this time to Greece.

The Ionian and Aegean seas held more islands and bays and channels than I could possibly explore in one summer, and maybe even two would not be enough. I liked Andraitx as a winter home, but it was too far from Greece to make a series of return trips a practical plan. So I said goodbye to Andraitx for a year or two, and I hoped to find its twin in Greek waters.

For this first summer of Greek sailing I was to have a Sicilian crew. Paolo and Costanza were coming, with a friend of theirs as well, a Sicilian girl whom I did not know. I was going to meet them in Palermo, sailing *Leiona* there on my own.

At the last minute a crew for the Palermo passage produced itself. Richard Gardner had written long ago to say that they could not come sailing this summer as they were involved in changing houses. Just before I was ready to leave Andraitx he wrote asking if Diana could come for an early month, bringing with her a sailing friend of theirs, Mary Cartwright.

I sailed from Andraitx in April with the two women as my crew, starting back along the path I followed eastward the year before. We moved up the east coast of Mallorca to Menorca, and from there had a three-day crossing to Sardinia, making San Pietro and Carloforte our landfall.

Once again, the sea was empty of ships. I had felt sure we would meet some, and hoped that my crew would not let any come too close while I was asleep, by day or by night. My stress on the need for waking me at once if anything strange was seen lost me some sleep on our first night out of Menorca. Mary woke me urgently.

'Keith, come quickly! There is a big red light straight ahead.'

I swung out of my bunk and into the cockpit. The red point of light was now clear as the red horn of a rising moon.

My sleep was broken the next night as well, but not by Diana or Mary. I was disturbed by a puzzling noise, a concert of squeaks and whistles which seemed to be all around me. At first I thought I was dreaming, but when I sat up I could still hear the twittering noises, sweet and high-pitched, coming from different directions in short, quick bursts, rising and falling in pitch and strength.

I called out to Diana:

'What is all this chirruping noise? Have we sailed into a flock of birds?'

'No, it is not birds. I think the noise is coming from dolphins talking to each other. Come and look.'

I went on deck. It was too dark to see anything clearly, but I thought I could just see the ripple of dolphins surfacing and hear the sigh of their breathing. The sweet twittering was still going on, but I did not hear it as strongly from the cockpit as from my bunk, where my ear was lower and close to the water. I made a mistake in trying to shine a light on the dolphins and they were frightened away. They came back the next night to give us another short concert.

At Carloforte we arrived on a saint's day to find all the flags out and the shops shut. Luckily Scottolino had seen us coming in and he met us on the shore. He took our shopping bags in his charge, and by hammering on the right doors and sending off boys on bicycles he loaded us with fresh bread, meat and fruit.

We called in at beautiful Malfatana and found it even more beautiful in the greenness of early summer. Then we took our departure and headed for Sicily. Palermo was our first port of call, and we reached it six days

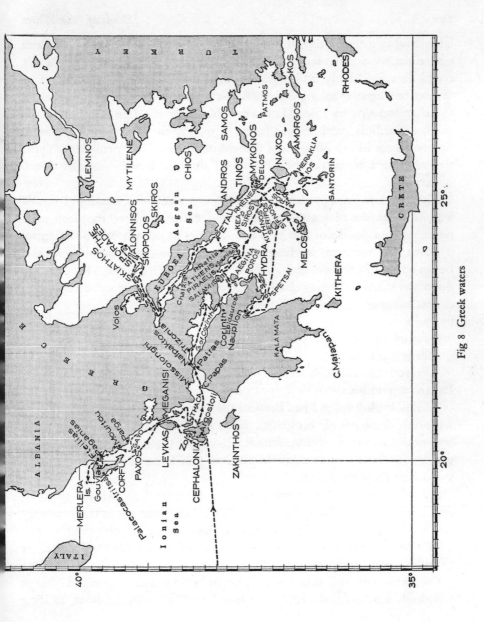

Fig 8 Greek waters

later, having met one day of bad weather. For the whole of one day and one night we were hove-to south of Sardinia. *Leiona* showed that she could lie comfortably under storm staysail alone. My only moment of dismay had been at midnight, when I saw two powerful red lights, close together and side by side. They were shown by the signal station at the south of Sardinia, safely to windward of us, and the only meaning I could find for them in my books was 'Storm or Strong Gale probable'. There was nothing I could do about it, and I thought it was best not to tell my crew.

The gale left us after dawn, leaving a big swell. I cooked and served a large breakfast of scrambled eggs and coffee, and we motored until noon, when a good north wind to help us on our way to Palermo.

Diana and Mary left for England from Palermo, and my Sicilian crew began to load in their clothes, gear and vast stock of food. This time the dock was shared with a cruise liner. For three days there was a repetition of last summer's social whirl, then we sailed, at midnight after a dinner party.

I found myself with a crew of four and not three. Costanza's first question had been:

'Do you not think it would be a good thing if we took one more Italian man with us?'

I did not think so, as I had become used to the comfort of sailing *Leiona* with only three people on board. Even four seemed too many, and five was going to mean overcrowding. However, Costanza is very good at getting her own way in a gentle manner, and I found that I had agreed to take on their extra man.

It was a crowd. Five people and their kit added to the pile of food, drink, masks and flippers, and all Paolo's fishing gear, made the boat very full, but we managed to live without discomfort or disagreement.

The four Sicilians had differing qualities which blended well together to make a good combination.

Costanza was the one who thought ahead and organized things winningly and unobtrusively. She loved just being in the boat, as long

1. At Zola: *r. to l.* Sandra, Costanza, Alexandria, Ioanis
2. Corfu

23. Petro's farmhouse, Corfu

24. Santorin, the Cyclades

as there was lots of time for swimming. Her only constant demands were for 'pretty places' and being woken for 'beautiful sunrises'.

The girl they brought with them, Sandra Ducrot, a Sicilian in spite of her name, was tall, dark, quiet and glamorous. She knew a lot about sailing and loved it. She steered better than any of us, or than anyone else who has sailed in *Leiona*, and she was happy to spend hours at the wheel. When she was not doing that she loved sleeping, so she was the ideal crew, competent and peaceful.

Paolo was a wonderful man to have on board, which was surprising because he did not really like or understand sailing. He became bored by it unless there was a strong wind, but he was always a good and willing helper, with weight and strength and courage. He loved spear-fishing and would be away for two or three hours, coming back with a bag full of large and small fish, and usually an octopus. Then he would spend hours in the galley cooking them. It was just as well that he liked cooking, at which he was very good, because neither of the girls ever cooked.

I was soon glad that Pietro de Capriolenti had been added to the crew. He became a great asset, even though he knew little about boats. Luckily, he wanted to learn, and he liked working. Unlike the others, he loved tidying things up, and on deck he saved me hours of folding and coiling and stowing. Best of all, he was a helper for Paolo, which was a very good thing, because Paolo preferred producing meals to clearing them up, and the girls preferred just eating them.

With plenty of watch-keepers on board we sailed straight from Palermo to Greece without stopping, heading for Argostoli in Cephalonia.

The passage along the north coast of Sicily was not at all like the slow, lazy tacking of the last summer. This time it was cold, but we had a good free wind to take us along fast.

The Strait of Messina greeted us with a swordfish, a rainstorm and a bat. The swordfish did a ballet of leaps close to us, darting straight up out of the water, high into the air, and falling back sideways with a slap and a splash.

The rainstorm enveloped us while we were still north of the Strait,

7

moving slowly towards it under staysail only, as we had arrived too early to go through. The rain came with lightning and strong wind and a fluttering bat. I first saw the bat as he was being blown past, trying hard to fight back against the wind. He just managed to grab hold of the rigging, and then to work his way crawling along one of the crosstrees. Then he hung on firmly upside down below the crosstree, clutching round the lee side of the mast with his spread wings. He hung here for half an hour and then tried to move. The wind whipped him off the mast and away to leeward. He tried hard to get back to the boat, going low over the water where the wind was not so strong. Then he dipped too low and the top of a wave caught him. He tried to fly off the water, but could not rise from it. I saw him swimming for a while, with his wings spread wide and his body held high. Then he disappeared.

The rain cleared as we entered the Strait, but the wind held and we ran south before it, setting and booming out the red genoa for an exciting sail with Sandra at the wheel. This time there were no fast boats out to chase the swordfish and the waters of the Strait were disturbed and white.

We turned east around the toe of Italy's boot and set a course for Cephalonia. At sunset we left the dark mountains of the toe astern, and for two hours moved across the lights of a busy steamer lane. Then we had the sea to ourselves until we reached Greece.

The crossing of the Ionian Sea brought no troubles or excitements, but there was enough wind to keep us sailing all the way, even at night. Paolo started to be bored after two days at sea, but luckily on our last morning we ran into shoals of small tuna, feeding and greedy. Paolo and Pietro got out their lines and soon hauled in four good fish. I was glad when Sandra persuaded them that they had caught enough. I thought we had enough fish to eat, and I was sure there was too much blood and scales on the deck.

Paolo was full of happiness in catching the fish and in getting them ready for eating. Before this excitement he had found time heavy on his hands, even though he always seemed to be making up something to eat or drink. When we left Palermo I had made out a watch list to last us for the summer, and had shown on it the times of breakfast, lunch, and supper. As soon as Paolo saw it he said:

'Please, can we make some changes in this? It is not nearly full enough. We must have many more social fixtures.'

The amended list was much fuller. It read:

a.m.	0.00–2.00	Paolo		
	2.00–4.00	Pietro		
	4.00–6.00	Keith		
	6.00–8.00	Keith		
			7.00 a.m.	Coffee
	8.00–11.00	Costanza	8.00 a.m.	Breakfast
			10.30 a.m.	Snacks
p.m.	11.00–2.00	Sandra		
			12.30 p.m.	Cocktails
			1.30 p.m.	Lunch
	2.00–5.00	Pietro		
			4.30 p.m.	Tea
	5.00–8.00	Keith		
			6.30 p.m.	Cocktails
	8.00–10.00	Costanza	8.00 p.m.	Dinner
	10.00–12.00	Sandra	10.00 p.m.	Nightcaps

Greece revealed itself to us majestically and beautifully. First we saw two pyramids of white cloud above the empty horizon, over Cephalonia ahead of us and over Zakinthos to its south. Then we saw the land below the clouds. We were closing on the south-west of Cephalonia. A mountain range lay across our front, climbing to more than five thousand feet. As we sailed nearer the land began to take shape and colour. It was greener than the high north coast of Sicily and more striking. The slopes were covered with woods, terraced with fields and orchards, and dotted with white villages. The steep coastline was formed of cliffs of white and red rock, many of them pierced by caves.

We rounded a point, sailing over shallow waters at the end of long, low fingers of shining white stone, and came into the long, narrow Livadi Bay, within which Argostoli hides in its own small bay on the eastern shore. By then it was evening and we did not want to go into a town harbour and undergo the complications of port officials at the end of our passage. We sailed on three miles into the long bay and anchored in its shallow western side.

Our introduction to Greece and its people was a good one. The entrance into Cephalonia had been beautiful and from our lonely anchorage we watched the colours of the sunset over the low hills and the next day's dawn over the mountains. My Sicilians were delighted by the act of the first Greek they met in his own country.

Before we had anchored in the evening a small boat had sailed near us. In it were a fisherman and his young son. He asked us if we wanted to buy any fish and looked disappointed at our refusal. Then Paolo asked him if he would like one of our fish and held up a tunny. He was very doubtful about the gift and smelled it to see if it was all right. Then he took it, thanked us, and sailed away.

As we were leaving in the morning we saw him coming back to us, from two or three miles down the bay. He waved and greeted us, thanked us again for the fish and gave the girls a box of Turkish Delight as a present of gratitude.

In Argostoli we came alongside the dock and a sailor from the port office boarded us with our mail and with a Transit Log for the summer. We filled up with fuel and water, did our shopping, and sailed out to a steep, narrow bay just beyond the harbour entrance. It was a good place for Paolo who speared a big *morena* eel which lashed about the deck biting pieces out of its own body in an effort to get rid of the spear.

Our Greek sailing started with a pleasant encounter. From Argostoli we sailed along the seaward coast of Cephalonia with a good wind to take us north. We had thought of going to the village harbour of Assos, hidden behind the walled peninsula of its Venetian fort, but we changed to the idea of finding a lonely bay where we could swim and fish. Assos was in the big Myrto Bay at the north-west corner of Cephalonia, ten miles wide. Right at the south of this big bay we saw an open cove which looked as if it would be a good anchorage in the day's fair weather. We could see two houses near the beach and a small village on the hillside above the cove. As we neared it a fishing boat motored past us, heading for the cove. We hailed them and were told in English that it was a safe place to anchor.

It was a good choice, with clear sand and weed for anchoring and with an arm of rock on either side for Paolo's fishing. He was soon back with a bag of fish. One of them increased the bag by disgorging on the deck an octopus and two crabs to decorate our evening platter.

Ioanis, the captain of the fishing boat, rowed out in a dinghy and invited us ashore for coffee. Paolo had gone fishing again and Pietro wanted to stay on board but Sandra, Costanza and I went with Ioanis. We had a delightful evening, first by the beach and then up in the village. No one lived on the shore, but a small breakwater had been built to shelter several open boats and the two houses we had seen formed a kitchen and a coffee shop for the fishermen. We met some of the villagers there and they asked us to go up the hill to see their houses.

It was a steep climb, fairly short by an old footpath but long by the way we were taken to see the new road being built. It was a village of only twenty houses and was called Zola, not shown on any of our charts. We bought bread and wine and were given baskets of figs and apricots. Ioanis took us to meet his pretty young bride, and an immediate friendship sprang up between his Greek Alexandria and our Sicilian Alessandra. It was dark when we returned to the beach, this time by the footpath, after promising that we would not sail away the next morning.

Ioanis came early to take Paolo away for a morning of fishing, and told us that we were all expected for a lunch cooked by his Alexandria, to be eaten at the house on the beach at noon. It was a good lunch of soup, fish, bread, wine and fruit, and we spent a happy day among the people from the village and the boats. We filled a milk bottle straight from a tethered goat, and were shown the raised beds, like children's tree houses, which were built in the branches of poles and brushwood for cool sleeping when the really hot months of summer came.

Our conversations were carried on in Italian, which some of the fishermen knew, English with Ioanis as the interpreter for the others, and Greek with the help of a dictionary and a phrase book. I had tried to learn a little of the new language during the winter, and I was pleased to find that some expensive years of classical education had not been wasted, once I could force myself into a completely different pronunciation.

In the afternoon we left Zola, and we felt that we were leaving friends.

We spent the month of May moving slowly north to Corfu, and then back to Argostoli before passing into the Aegean sea. We could happily have stayed much longer. The islands were green and gentle below their high peaks, without the white glare which we were to find later in the more barren Aegean islands. The winds, too, were gentle, blowing to a pattern of light southerly airs in the mornings and changing to a stronger north-west wind in the afternoon.

All the days and all the places were beautiful, and different. Of them all, our favourite sailing ground was in what we called the Meganisi Lake, among the islands scattered in the lovely triangle between Levkas and the mainland, with the island of Meganisi as the centre point. Ithaca and Cephalonia lay to the south, and the high pyramid peak of Levkas was on the north-west. To eastward the hilly coastline of the mainland rose into a mountain mass which swept northward beneath a pile of white cloud. In the afternoons these clouds would change into black anvils and we could see the flicker of lightning and later the wide flashes bright between the dark of the clouds and the mountains. These thunderstorms stayed far inland and all that reached us was the steady wind of the afternoon and the calm of the night.

We worked northward very slowly, finding beautiful places but meeting few people. Only in the deeply hidden village harbour of Kioni, at the north end of Ithaca, did we find ourselves adopted into the life ashore. It was on a feast day, and the village was bright with lights and music.

We went ashore to fill our shopping baskets and were taken under the care of two Greeks who had come back to live again in their village after years abroad, one from South Africa and one from Australia. All the shops were shut, but we were led through side lanes, back doors and gardens until we had all we needed.

Our shopping took a long time and we had been entertained by half a dozen families in their houses or courtyards before it was over. Then we put our baskets and wine flasks under a tree in the village square opening on to the harbour and we entertained our guides. By then it was nearly midnight and the music and dancing were in full swing. The square was full of trestle tables, each one more than filled with a family group of all ages. A six-man band played without stopping, mostly tunes of modern dances but sometimes changing to one of the old national airs, even more popular than the modern dances, and bringing the old people on to the dance floor. The families at the tables clapped and laughed and shouted, and joined in the songs of the country tunes. The local dances were pretty to watch, with their advancing and retreating lines, the complicated steps and patterns, and the swirl of the skirts and scarves and hand-kerchiefs.

The trestle tables were filled with food as well as people, and the air was heavy from the blue smoke of grilling meat. A grill had been set up in the courtyard of a taverna and a team of cooks worked away all night. There were some waiters, but not enough, and most of the diners went to get their food when they wanted it, which was often. There was no attempt to provide plates or knives and forks, and the piles of blackened lamb or goats meat, potatoes, onions and bread were served on folded sheets of newspaper. After that fingers and pocket-knives came into use. It was a feast of quantity, not of quality, and could well have been called Homeric, set in the island of Ulysses.

The Kioni feast was our only experience of social life on our Meganisi Lake visit. For the rest of the time we hardly saw another person except when we walked or sailed into a village to buy fruit, wine and bread. Nearly all our anchorages were in deserted bays or in small coves within a bay. There was seldom room to swing at anchor and one of us would swim ashore to fasten a stern-line to an olive tree. Sometimes we did not use an anchor at all and moored both bow and stern to the olives, in what seemed to follow a good classical tradition.

We did not want to leave our lake, with its lovely bays and islands, olive groves and sheltered waters, but the time had come to move on to Corfu, a hundred miles to the north. We motored through the marshes of the Levkas Canal, passing close to the Venetian forts at either end, and were once more in the open sea.

After months of clear, bright skies, it was a shock to meet a hazy, almost foggy day. The coastline was steep and high, yet we could not see it from two miles at sea. This haze hung for the whole morning, with a very light on-shore breeze. In the afternoon the wind got up a little and the haze lifted slightly, but never quite left the hills.

A sign on the chart showing a freshwater whirlpool in the middle of a deep bay attracted us. We sailed into St Giovanni Bay, found the whirlpool, and spun gently around in it while we drew up water in a weighted bucket. With effort we could imagine the water to be a little less salt than the sea, but I think that a very deep-reaching pipe would be needed for any success.

When at last we came to Corfu, we enjoyed the city but not the harbour. The approach was beautiful, sailing towards the rocky peninsula of cliffs and the fortress walls of the citadel, joined by a narrow strip to the white houses and dark green trees of the city, with the high hills beyond across the bay.

It was fun to walk along the broad streets and open squares and parks, forgetting *Leiona* and being real trippers, entranced by the kilted Evzone sentries in front of the palace. In the heart of the city were the narrow winding alleyways that we had become used to seeing. I was led through a maze of these in what I thought would be a simple search for some distilled water.

I took the half-empty bottle ashore to fill it, and went to a garage, which seemed the obvious place. The first garage had none but the owner gave me some directions which took me away from the waterfront and into the narrow streets. I still thought garages were the proper sources of distilled water and tried two more without success. Then I remembered that in Spain I had once asked at a garage and had been directed to a chemist's shop. So I tried three of them without getting either the water or advice on where to find it.

By this time I was far off the beaten track of tourists and I could find no one who spoke English or French. I had not thought of trying Italian, and my little knowledge of Greek was not helping me to trace the apparently rare thing I wanted. I was beginning to feel lost in my efforts and I was certainly lost in the winding back streets. Then the inevitable Good Samaritan turned up, a man who knew some English and was able to follow my poor attempts at speaking Greek. At first I thought that he only pretended to understand me, because he took me straightaway to a tailor's shop.

I kept showing my bottle to him and to the tailor, trying to explain that it was water and not clothes that I wanted. I was reaching the final stage of the Englishman abroad in difficulties, beginning to talk slowly and loudly, but my guide assured me with voice and waving arms that all was well. There was no water in the first two shops, but in the third one we met success. Instead of being turned away, we were led through a forest of hanging suits into the family parlour, to the surprise of the many daughters, aunts and cousins sewing there. From there we went up a narrow staircase between two walls and came to an upper room.

In front of us was an ironing-board and a man working at a huge iron. This iron was connected by pipes and tubes to a big kettle of water steaming away over a coal stove in the corner. Other tubes led upward to the ceiling and then down to a glass jar into which were falling, at last, clear drops of distilled water.

The harbour of Corfu was dirty and uncomfortable and even dangerous, at least for a small yacht. We moved out to Gouvia, a lagoon-like bay five miles to the north. Here we anchored near a little chapel on a causeway, in front of a low white farmhouse half hidden by plane trees and honeysuckle. The farmer and his wife, Petros and Antigone, and their three daughters were as charming as their house. They did our laundry and mending, sold us eggs and bread, gave us milk warm from the cow, and brought us baskets of fruit and flowers.

Gouvia was a good base for Corfu which was quick to reach by bus or taxi. There were pleasant walks around the pretty lagoon, where the only jarring note was struck by a new hotel on the western side. But the hotel was not as unsightly as many and it offered a sophisticated change from the simple life provided by Petros on the opposite shore.

Cephalonia, our first landfall, had given us the pleasant adventure of meeting Ioanis and Alexandria in their village. Corfu was our turning point, and there we had an adventure which could have been an unpleasant one. It was my fault for not being more careful and inquisitive.

On our way to Corfu we had found two uninhabited and well-sheltered bays on the mainland coast opposite. This was after we had been anchored in the pretty channel between Sivota and Mourtou, close under the cliff-top village of friendly people, I had asked Christos, one of the fishermen, if he knew Paganias, the first of the two bays. He looked startled, shook the fingers of one hand, and said something about the 'bad Albanians'. English and Greek charts and maps showed Paganias to be in Greece, although close to the Albanian frontier, and the *Mediterranean Pilot* told of a Customs House there, so we thought that we could ask the local authorities for more information.

We found Paganias completely deserted. It was a small, land-locked bay and seemed like a mountain lake. The reed-fringed water was still and dark green, within the surrounding slopes of bare hills covered with grey rocks, short, faded turf and a few bushes. Above the entrance was an empty hut, and at the southern end was a ruin in the water and a tumbledown stone house near three reed huts, shaped like beehives. There was a sinister feeling of mystery, unlike anything we had met. Its atmosphere was saved from complete gloom by the strong scent of honeysuckle and the singing of larks.

Early in the morning we left, gladly, and sailed slowly along the bays and hillsides of the coast to our next choice, another landlocked bay called Ormos Fteilias. This was a change from Paganias and was gay and pretty. Pietro did not want to go there as he was sure it was in Albania. Again our charts and tourist maps showed it close to the frontier, but apparently just inside Greece. It looked so pretty that we entered, sailing into a deep, narrow arm, hidden from the sea. We were reassured by seeing painted on a rock in large letters '*Landfall–1945*', and we felt that the place must be all right if this American yacht had been able to stay there. We walked and swam until the afternoon and then sailed for Gouvia, meaning to return soon to Fteilias.

We returned a few days later and at once ran into trouble. As we neared Fteilias, which is only five miles from Corfu, we noticed a black vessel like a small trawler moving towards the bay, having come from the north, where the coastline was certainly Albanian. A white flag flew from the gaff of her single mast. She went into the mouth of the bay, turned with her bow almost touching the western point, and then went back the way she had come. We were four or five hundred yards away and could see that her flag was an ensign with the double-headed eagle of Albania, and that there was a gun on her foredeck. I was glad they had taken no interest in us. They disappeared astern and we entered the bay, turning eastward down the arm we had visited before.

At the head of the bay we anchored and began to think about tea, swimming, and fishing. Then I heard a thudding engine and saw a Greek fishing boat coming fast towards us from the mouth of the bay, with excited men waving in the bows. I thought that they were going to hit us hard. One man ran aft, possibly to put the engine astern, and they slowed down, but even so touched us hard enough to take off some paint. There were eight or nine men on board and they all spoke at once, shouting and gesticulating about bad Albanians, shootings, handcuffs and danger. They urged and beckoned us to follow them out, keeping close to the eastern shore.

Outside the bay the urgency and excitement lessened and they beckoned us alongside. I put our bow near their stern and two men jumped on

board. I thanked them for getting us out of trouble, and we shared our
tea with them. They told us they were soldiers, although no part of their
clothing was military uniform except possibly their brown pullovers.
They came from an army post halfway to Paganias, which we had seen—
a cluster of white huts with a flagstaff and a wooden jetty. When tea was
finished I suggested putting them back on board the fishing boat so that
we could go on to Paganias for the night. This suggestion was not well
received and they insisted that we must come to the post and wait there.
I then realized that we were under arrest.

At the post I let go a stern anchor and laid the bow near the quay.
Things were getting beyond the scope of my scanty Greek and an 'interpreter' was produced, a tall young soldier who brought with him a phrase
book from which he was teaching himself French. He did not help us
much, because he could only transmit and not receive. He spoke fluently
the few phrases he had learned like a parrot but he could not understand
anything said to him.

There were about twenty-five soldiers at the post with only a corporal
in charge that afternoon. He had to send by wireless for the lieutenant
or sergeant to come back.

I could see that we would have to stay the night and that the whole
case would work its way up to higher levels and might take a long time.
The immediate action was to put the boat into a safe place, as a swell was
starting and I felt her touch the bottom. I explained this and moved out
to anchor two hundred yards off shore. Luckily for our comfort, they
did not think in time of putting a guard on board. The female opinion
of my crew was that we should motor off in the dark, but it was obviously
a bad idea and could only lead to much more trouble.

In the morning I could see that a sergeant was at the post and I hoped,
but doubted, that he would have authority to let us go. I shouted to ask
if we could sail and was met with a chorus telling us to 'stay here'. It was
no good carrying on at shouting distance so I started to move closer. I
wanted to see what their reactions would have been to an escape so when
the anchor was clear I circled away from the jetty. The reaction was very

clear. Four men ran off to the huts and came running back stuffing cartridges into the magazines of their rifles. Luckily we were heading towards them before they could start shooting.

We had to wait until a Customs launch arrived with an officer whom I took to be a lieutenant in the Intelligence Corps. He spoke nothing but Greek, so we did not learn much except that we were to follow him to where the colonel was, at Sagiada, south of Paganias. We anchored there outside the shallow harbour and waited for developments. These took the form of the arrival from Corfu of a naval frigate. The commander went ashore for a long conference and then came out to *Leiona*, bringing the lieutenant with him. He spoke perfect English and was very friendly, even though he was wearing a pistol.

We settled down in the saloon with the charts, the log, and a bottle of whisky and went over the story, which was translated for the army officer to write down. After an hour, the commander sat back and said happily:

'Good, at last all that is over. What a lot of trouble about nothing.'

We agreed with him, filled the glasses and drank healths all round. The commander sat back again, and with the same happy smile and voice said:

'And now I must ask you to follow me to Corfu. Naval Headquarters need to hear your story.'

There was nothing else we could do, so back we went, this time into Corfu harbour. I put on my best clothes and was taken to the naval officer commanding Corfu. He was charming and sympathetic, and apologized for all the inconvenience. I apologized for all the trouble I had caused.

He showed me his marked charts, which were English. We had not crossed the Albanian frontier, which ran through the middle of Ormos Fteilias, but we had sailed through Albanian waters near the headland. We had also sailed many miles in Greek forbidden water, which included Paganias, although we had no way of knowing it. I was glad that we had been arrested by the Greeks and not by the Albanians. It had been a lucky escape.

The next day we started south again, keeping well away from Albania. First we went north around Corfu and then down the seaward side, anchoring in two wide sandy bays until we came to the delightful village of Palaeocastritsa. Here we lay in a crater-like bay between rock walls, looking almost straight up at the ridge above us with its mountain village and monastery.

A strange morning squall blew us awake and out of the bay before sunrise, and we ran under staysail down to the mouth of the little river at Ermones. We had met a rival claimant two days before, but this surely was the place where Ulysses met Nausicaa and her maidens. The cascading basins of rock are there, and the day we came in the maidens were there as well, washing clothes and laying them out to dry on the bushes and hot stones.

From Corfu we retraced our way south, finding some new places and going back to others that we had liked. The village square of Gayo and the castle of Parga were hard to leave, and we longed for more time. We hated to leave our Meganisi lake, and the golden bay we found at Anti-Samos on Cephalonia, where in three days we saw only one man, and then only for five minutes. We wanted to see Ioanis again, but we could not. We had made the usual sailing mistake of planning to go too far in too little time.

I would have been happy to stay all summer in the Ionian islands, but the temptation to see more of Greece was too strong. We wanted to see the Cyclades in this summer, and there was not much time left if we were going to avoid the strong midsummer winds. I had been told how uncomfortable and at times impossible, sailing in the Cyclades could be when the Meltemi, the strong northerly wind, was blowing. I gathered that we might meet the Meltemi at any time during the summer in the Aegean, but that until about mid-July the blows would be short ones. From then until well into September we could be certain of lots of Meltemi, sometimes lasting for several days at a time.

My Sicilians were staying with me until August, so we had only June for our Cyclades wandering, and some of July. When the Meltemi began to become too strong for comfort we could leave the Cyclades and look for more sheltered water in the big gulfs south-west of Athens.

So at the beginning of June we left our beloved Ionian waters and sailed towards Corinth.

CHAPTER SIXTEEN

To Corinth and the Cyclades

The golden bay in Cephalonia was our last Ionian anchorage. Then we sailed east, into the narrow gulfs of Patras and Corinth. Our first approach to the land brought us a shock. We entered the Gulf of Patras in the evening and anchored for the night off the sandy beach just inside Cape Papas, at the mouth of the Gulf.

A small fishing boat was passing near us, and we asked him for some fish. He answered:

'I have none now, but I hope to catch some soon. Where will you be at eleven o'clock?'

Paolo wanted fresh fish for his galley, so he called back:

'We will be here, with a light showing.'

The fisherman looked surprised.

'You cannot stay here. This part is for military only. They shoot at anyone who goes near the land. You should go into the harbour and see the captain.'

With memories of Fteilias still fresh, we did not want to get into trouble, but nor did we want to move. We were nearly half a mile from the shore, and near a main steamer route. Besides, supper was being cooked. We decided to stay and see if anything happened.

Something did happen.

Just after dark I heard what I thought was a rifle shot, but I was not sure. We all looked and listened, but there were no shouts or lights or whistles. I might have been imagining things.

A few minutes later there was the 'crack' of a bullet close overhead, and the 'pop' of a rifle on the shore. And then the same again. This time there was no doubt. We switched on all the lights we could to show that we were moving, got up the anchor, and motored across to the other side of the gulf, to anchor in the shallows on the Missolonghi shore.

Paolo was silent for a while, and then said:

'Keith, you may have left the army, but the army doesn't seem to have left you. I wonder when we will have our third meeting with rifles and bullets and threats of war.'

We hoped that we had left trouble behind. A west wind took us through the narrows of Patras and down the Gulf of Corinth. It was perfect sailing in lovely scenery. We had blue water around us and high mountains on either side. The valleys and long slopes of the hillsides were green and silver with plane trees, pines, and olives, with white villages and red or brown fields breaking the colour.

We made two stops in the Gulf of Corinth, close to each other, but in very different settings, and both delightful. These were in Nafpaktos and Trizonia.

Nafpaktos was the smallest of harbours, and is more widely known by its old Venetian name of Lepanto and the naval battle named after it. This was the last galley battle of history, when three hundred Turkish ships were defeated in 1571 by a fleet from Venice, Genoa, and Spain, under the command of Don John of Austria.

We could see no signs of boats or harbours as we sailed towards the town. Even when we were close enough to smell the orange blossom all we could see were houses grouped close together at the foot of a steep hill, and in front of them a high sea wall. We sailed closer. Then we saw how the harbour was formed. On the top of the hill was an old fortress and from the fort the walls ran down along two knife-edge ridges. Near the shore the walls curved inward to enclose the town. They did not quite meet. In front of the centre of the town they left the land and continued their sweep into the water, still curving towards each other. Then each wall ended in a squat walled tower, leaving an open gateway only forty feet wide. Through this entrance, between the walls and the houses, was the harbour, a round bowl seventy yards across. Inside were some small open fishing boats. There was just room for us.

The stern lines were made fast near the foot of a big plane tree, and we sat under its shade, with our evening glasses of *ouzo*. The tree was a home for sparrow hawks and scores of them were darting in and out of its branches and circling above it.

The next day we exchanged the close circle of Lepanto's walls for a ring of vineyards and olive groves within the low hills of Trizonia's small island. The little village was on the north of the bay, astride a low, narrow neck. Over the houses showed the masts of the fishing boats at their summer moorings on the seaward side of the village, and we had the bay to ourselves.

Above the masts the mountains made a magnificent backdrop to the scene. On the lower slopes were the greens and browns and pale golds of the fields and vineyards. Higher up the olive groves made a silver-grey carpet, marked by the dark pillars of cypress trees. Then came the pine woods, mostly dark, but with patches of light green. The dry water-courses made huge triangles of red and pink where the oleander was flowering, starting as points among the pines or olives and spreading upwards to cover acres and miles of the hillsides.

At the end of the Gulf we left *Leiona* for half a day in the little harbour of Corinth, and went off to see the museum and ruins of Old Corinth, and to clamber up the hill to the fortress above it. From there we had a splendid view of the gulfs behind us and of the waters and islands ahead, which would lead us into the Aegean Sea.

That afternoon we went through the Corinth Canal.

As we sailed down to its head a red flag was hoisted on the shore. We thought it was to stop us because a ship was coming the other way, but it was because of our sails. As soon as we had lowered them for anchoring a blue flag went up, and two sailors waved us forward. The following wind was too good to miss, and there was no ship astern of us, so once inside the canal we got the sails up again and went through in silence.

I had expected the canal to be bigger, but even so it gave us an impressive sail. It ran for three miles as a straight thin cut, between steep sides of stone. Halfway two bridges produced a scene like a picture in a children's book illustrating different kinds of transport. As we passed under the bridges we could see overhead an aeroplane, some cars, people walking, a mule cart, and a train. There were also small boys dropping stones into the water. From nearly two hundred feet the stones fell with a big splash, and we hoped that our deck would not be too tempting a target.

At the far end a launch came to collect the canal fee, a heavy one of about £7, which was the minimum, however small the boat. We sailed on until dark, and anchored in an open bay on the southern side of the Island of Salamis.

We had to go to Piraeus to collect our mail, but we wanted to leave as soon as possible and start our sailing among the small islands of the Aegean.

Because Piraeus was the port of Athens I had in my mind a picture with Athens as the dominating feature, and I could imagine sailing in towards the Acropolis and the Parthenon. If we had been coming from the south we would have seen them, but from the west the first signs of a big city were smoky factory chimneys.

I lost my way that morning and arrived in the wrong harbour. There are three harbours at Piraeus. On the north side of the town is the big commercial harbour, Piraeus proper. On the other side are two small yacht harbours, officially called Zea and Munychia, but usually spoken of by their old names of Passalimani and Turkolimano. We wanted to go to the Royal Hellenic Yacht Club in Turkolimano.

Close to the town we got our sails down and motored on. My chart showed a breakwater at Turkolimano and none at Passalimani. We turned boldly in at the first breakwater we came to, and after making fast and filling in some papers we found that we were in Passalimani. My chart was out of date, and I had not learned about the new breakwater. I was very apologetic to my crew as we left the wrong harbour and found the right one.

Piraeus was hot and dusty, and we were glad to leave in the afternoon. That night we lay at anchor below the temple of Poseidon at Sounion. We were beside a rocky island which was the home of thousands of swallows. The view from the temple at sunset was wonderful, and it was even better at dawn when I stood there alone in the silence. To the north and west were the mountains of the mainland. South-east lay the nearest islands of the Cyclades, black hills in a silver sea, slowly changed by the rising sun into golden islands in blue water.

When I think of the Cyclades my mind has a picture of gleaming white islands scattered over a bright blue sea. This is not a true picture, as their colours are more light brown and yellow, but the impression is of whiteness, the shining white of the houses and churches at the waterfronts, and of the beaches and cliffs of the shores. The deep water was always blue for us, and it was only over the shallows close in-shore that it turned purple above weeded rocks, or pale green and white over the clear sand.

We had five weeks of wonderful sailing in the Cyclades, and we wished that we could have had five months. The weather was perfect until July began to show its warning of the Meltemi season. It was a delight to find so many islands and bays and beaches in so small an area. In a space only a hundred miles across there were twenty major islands and many more small ones. Wherever we sailed we could see land in some direction and find a good anchorage within a few hours. It did not matter what wind each morning brought because we could always find some new place or a favourite old one on a comfortable course to leeward.

From Sounion we sailed south-east through Keas and Kithnos to a big and safe bay at the south-west corner of Siros. We stayed there for three days and then started for Delos. Of all the places in the Aegean, Delos was the one we most wanted to see, and we decided to make it our first port of call. The wind changed our minds. When we cleared Siros we found a strong wind blowing from the north-east, almost from Delos.

We could have got to Delos, but the passage would have been slow, wet, and uncomfortable, so we went somewhere else and came back to Delos later.

We went south, making the weird volcano of Santorin our final target, unless the wind should change our minds again. Our southerly route took us by way of Paros, Naxos and Ios. It was a pleasant journey.

Paroekia, the port and capital of Paros, was our first town in the Cyclades and we were delighted with it. There were few streets in the town, only paved passageways between close-packed white houses. The stairways of the houses were on the outside, angled or curving as they led to the upper storeys or flat roofs. Haphazard among the houses were churches or small open triangles or paved squares, always with a bar or coffee shop on one side. Thick vines of honeysuckle draped over the white walls and scented the air. The steps of the houses and even the edges of the passageways were whitewashed and the glare in the bright sunlight was intense.

The people were friendly without being clinging, and interested but not prying. With phrase-book Greek, signs and the help of an occasional person who spoke English or American, we found that we could find our way and do our shopping and even get information. We found one man who spoke French, but Italian was no longer any use. These islanders started their days early and people were moving about on the quays as soon as it was light. Between 1 p.m. and 5 p.m. was the quiet time of the day, when all shops shut, to open again in the evening and stay open as long as there was any trade.

I filled with water and diesel fuel at the quay, and found what good arrangements were being made for yachts. In many small ports part of the quay had been marked with blue and yellow stripes as safe for mooring a yacht. At most of these places there was water and duty-free fuel, which could be piped straight into the boat. These arrangements applied to some seventy ports in Greek waters, ten of them being in the Cyclades. It was a great help not to have to drag water cans to village pumps or roll dirty fuel drums to the boat. The only snag with the marked quays was the uncertainty of depth, which was supposed to be at least three metres, but was often much less.

We spent another day in Paroekia, and then started our wandering southward. The days were all alike, and the pattern of our life and sailing changed very little. Only the places were different, within the constant warmth and beauty of the Cyclades.

Paolo and I were the early risers. He would start the day by making coffee and getting the breakfast ready. If we had followed our usual habit of anchoring in a lonely bay, I would go off swimming. Then I would start getting the boat ready for sailing or putting on the awnings if we were not moving that day. By this time the others were beginning to stir. They all liked their rest, and only surfaced when Paolo's clattering in the galley and saloon and my tramplings on deck made sleep impossible. Costanza might have watched a 'beautiful sunrise', but she would have gone back to sleep again.

When everything was washed and cleared away we would sail at leisure, or go ashore for walking or shopping, or sail the dinghy and swim. If we were moving that day, we would try to find an anchorage early in the afternoon in time for a swim and a wander ashore before cocktails and supper. We went to sleep early. It was a pleasing life, and although we apparently did nothing, we were never bored.

We called at Ios and then had a fast sail south, watching Santorin rise out of an empty horizon.

It looked like a round hill whose top had been sliced off flat. It was the peak of an old volcano, forming an island ten miles across, a circle of rock enclosing a seven-mile crater, which was the harbour. The circle had broken away in two places, leaving a narrow entrance in the north and a wide one in the west.

We sailed into the big crater through the northern entrance, between cliffs of white and red. The seaward slopes of the island were gentle and fertile, covered by fields and vineyards. Inside the crater there were no slopes at all. All around us were almost vertical faces of dark-coloured rock, nearly a thousand feet above our heads. The highest point of the island was a peak of nearly two thousand feet, but we could not see it over the cliffs which shut us in. Near the middle of the crater was a cluster of small islands. A recent eruption had extended one island eastward by a long apron of black lava, looking like a slag heap. The village of Thira was high on the crater's edge, on the eastern side, facing the lava island. We saw a small quay with a few houses and open boats at the bottom of the cliff below the village and we sailed towards it.

As we neared the quay two men rowed out to meet us. They said that the place was 'not good', and pointed to an iron mooring buoy, big enough for a steamer, two hundred yards away. We gybed round and went to it. The men took a line and made us fast to the buoy. The depths inside the crater were too great for anchoring. In the clear water we could see the thick chain of the mooring going down and down until it disappeared into the dark water more than a hundred fathoms deep. We paid harbour dues of one shilling for the use of the buoy.

The whole crater was an amazing sight, and the village was fantastic. It seemed to be straight above our heads. Along the top of an eight hundred foot cliff of red, brown and black rock ran a silhouette of white houses, some with flat roofs, some with domes and arches. The straight and curving lines were broken by the points of spires and belfries. Only the tops of the buildings could be seen and there were no signs of people or of movement.

Between the village and the quay the dark cliff was marked by the straight lines and sharp corners of a white stairway which zigzagged up the shortest and steepest route. We walked up those stairs in the afternoon. It was a long climb. We could have ridden up on mules, but the climb was a challenge, and the mules looked tired enough even without our weights to carry.

Because of its depth of water, Santorin is a favourite port of call for the big cruise ships and the muleteers are busy when the tourists flood the island. The starting price given us was fifteen shillings for each mule. When we said that we wanted to walk, the drivers did not seem unduly disappointed, and followed us on our climb, hoping that we would weaken. In any case, it was time for them to go up to their homes in the village. They were persistent in their offers of a ride. The interesting thing was that as we climbed higher, so did the price of the mules, until very near the top, when the market fell rapidly. We managed to resist temptation.

The stairway was enclosed by stone and plaster walls, newly white-washed. We walked on cobbles, but an exotic touch was added by the facings of the stairs, made of marble slabs. That was the only exotic thing about them. There were no sweet perfumes of orange blossom or honey-suckle, only the more earthy smells of sweating men and beasts and mule dung.

When at last we reached the top, we found Thira bigger than the silhouette had indicated. There were shops and restaurants and two big churches. The cruise ships had created a demand for a tourist office and a studio of paintings, statues and local crafts. Many of the houses were not lived in, because they had been damaged to a point of danger in the last earthquake a few years before. There were still signs of life in the heart of the volcano, and sulphur was bubbling beside the lava-island. No one wanted to live by the quay in case an earthquake or eruption brought the cliff down on them. The double journey of the steps was worth the feeling of a safe night's sleep.

Our evening walk down the steps was cooler than the climb. The sunset was beautiful, the last rays outlining the western walls of the crater and falling through the gap to make the quiet water of the harbour mauve and orange and opal. As the sky grew dark the scattered lights in the houses above us mixed with the stars over the black curtain of the cliff.

During the night a trading schooner had come to share the buoy with us, lying astern of us except when we drifted together in flat calms. She had used two rough, thick warps which hung far down into the water for most of the time. In the morning one short and solitary strong puff of wind stretched us both away from the buoy. The schooner's warps tightened and rose through the water, coming up on each side of *Leiona* and holding her firmly, taking off a lot of anti-fouling paint until we got the two boats clear of each other.

It was a hot, still day inside the crater, with little happening except the loading of pumice. The schooner which had shared the buoy moved off to lie near the quay, where she was joined by another. They were both a bright, orange colour which was a favourite in the islands. It looked just like priming paint, and maybe it was. It was one of the questions which I meant to ask, but never remembered at the right times.

The pumice was being brought down in the panniers of mules and donkeys, walking free and unled, a man driving each group of four or five with a stick, stones or shouts. The little brown and grey figures on the white stairway looked as if they were part of the feeding belt of some huge machine as they wound and turned down the angles of the track, making a pattern against the counter-flow of the empty panniers being carried up for another load.

At the quayside the panniers were swung off the donkeys into boats and rowed out to the schooners. There they were hauled up on tackles working from the masts, the pumice being emptied into the holds, and the empty panniers sent back to the shore. Loose pieces of pumice fell and floated on the water, and dust spread everywhere. Soon the two orange schooners were standing in a sheet of floating grey pumice, looking as if they were embedded in concrete.

From Santorin we sailed north and for ten days had good weather, sailing among small islands without a programme and anchoring in deserted bays. Each night was spent in a different bay, with leisurely sailing in between. We went first to Ios and then idled around the small islands east of Ios and south of Naxos, Heraklia, Echinosa and the Kupho Islands. Then to Paros, Naxos, Paros again and finally Delos.

There were many happy memories. Ios gave us two surprises. We arrived at its southern coast on a very hot afternoon, longing to anchor and swim. A little bay named Pikra Nera—Bitter Waters—looked perfect, and it was, in spite of the first surprise. At the mouth of the bay Sandra cried out:

'Here is a dolphin, leading us in!'

Sure enough, there was a black fin dead ahead of us, going our way.
I wondered if this was a local Pelorus Jack. It was not. The fin stayed
steady, moving forward but not diving. It was a shark, having a lazy day,
the only live shark I had seen in the Mediterranean. He turned before
entering the bay and swam along the shore line. We could see his tail break
the surface, showing his size, eight or nine feet long. I started the engine
and tried to close with him. He sounded and then came up again, a little
further to sea. Then he turned straight out to sea and disappeared.

We were sorry to leave the little bay. Paolo had grilled an enormous
steak over a scented brushwood fire and we had drunk our wine until
late in the night on the beach of sand and warm rock. This was a bay of
birds and bird song. There were ravens and grey crows, bright-coloured
flycatchers and sweetly whistling blackbirds. There was an unseen covey
of partridges chattering in the bushes and big, green grasshoppers, almost
as big as birds, flying slowly on wide wings, as well as lizards, green, grey
or brown, basking on the rocks, and darting under the leaves.

In the afternoon we moved round to the harbour and village of Ios
and had our second surprise. A motor cruiser from Scotland was stern-to
on the quay. As we were putting our dinghy over the side we heard a
strange noise from the shore. The noise went on and grew louder, a rough,
humming sound. Then the note changed, rose higher and became some
sort of a tune. It was the sound of bagpipes.

We were amazed. We were even more amazed when round the deck-
house of the motor cruiser appeared the piper, in full, proud music,
wearing a white shirt and blue trousers. He marched down the gangplank
and along the quay. Heads popped out from every window and the
villagers came out of their doors. Children flocked around and followed
him like the Pied Piper. He went straight to the arches of a taverna at
the far end of the quay and marched up and down in front of it, piping
strongly. By this time all the village was thronging round him. Then the
play began. From the taverna came four people—two couples, middle-
aged, short and soberly dressed. They stood near the piper and each

bowed to each. Then the ladies took their escorts' arms and they formed a procession with the piper leading. He piped them across the front of the village and aboard their ship for dinner. The villagers applauded and cheered and so did we.

Ios had yet another discovery for us. We had already met the Greek *mezes*, the olives or small pieces of meat or cheese that are served with the drinks before a meal. Here we had our first introduction to grilled octopus tentacles. They were excellent, grilled over charcoal and aromatic twigs in a brazier beside the table until they were crisp and savoury.

Pietro was a historian and liked searching for traces of the ancient ways of living. The place he enjoyed best of all was one we found by chance. It was a lonely bay on the east side of Heraklia Island, where a valley led into the hills. It was very small, and no one lived there. We saw from the chart that there was a village over the high ridge to the north, but in our valley there were only two fields, a few fir trees and palms, some wild olives and thorn and gorse bushes. In the centre of the valley stood a small, steep hill, a rocky bump shaped like a miniature volcano. On it we could see the deserted ruins of a walled village.

It was a fascinating village. What was left of the walls and houses was of stone. Some stones had collapsed and some had been taken away. Where the walls were standing the stone roofs were still intact. We found two bakeries, and these had been used recently. On a little mound within the walls, we found a big slab of marble in which a shallow basin had been carved, and a drain cut in one side. It may have been for crushing olives, but we liked to think it was the sacrificial stone of an altar.

We spent the day ashore, picnicking under the olive trees, and swimming and sleeping until the evening. Then we walked over the ridge to the village, Agios Georgios. On the way we met and greeted an old shepherd who surprised us by speaking English and telling us that he had fought in Italy with General Freyburg's New Zealanders. He became our interpreter.

The village and its people were charming. The houses were huddled close together on the two hills at the top of a steep valley. A track ran down to a narrow little bay with a small jetty. There were two new churches, but with the old seats taken from the former ones. Oddly enough, we saw no priests, whose robes, tall hats and long beards had become so much a part of the Greek scene that their absence was striking.

We were shown all over the village and then taken into the front room of a house and given orange liqueurs to drink. Most of the village people managed to crowd in and stand around the walls. Nearly every older woman held a distaff and its wool, and twirled it the whole time. It was a happy and friendly afternoon, and we felt very peaceful as we walked away between the white houses with their gardens of geraniums and clove orchids.

Paros and Naxos gave us some other beautiful anchorages, but no village was as delightful as that one on Heraklia.

Mykonos and Rhenea are close together and narrow Delos lies in the strait between them. We anchored one evening in Rhenea, the westerly island, in a bay facing the sunset. It held the clearest water I had ever seen. We lay in a little cove to the north. The water shoaled blue, green and turquoise over the smooth sand to a white beach. The cove was flanked by low sheer rocks, dark green and purple below water, white and pale grey above. On the low ridge at the head of the cove there were two stone huts and a donkey under a fig tree.

Before it was dark we all walked over the ridge to the far side of Rhenea to look at Delos and Mykonos. On our way back, carrying bunches of wild flowers, we met a very young and very pretty little girl. Her name was Maia and she had fair hair and blue eyes. She greeted us as if we were old friends and burst into an excited mixture of words, laughter, sounds and gestures. Her story seemed thrilling and fascinating, and we wished that we could understand what it was all about. We joined in her laughs and expressions of amazement, and encouraged by our support, her story began to reach its climax. For a time it seemed to be about a fire and an explosion, which startled us all. At the top of the ridge, we were relieved to see *Leiona* still safe in the cove. Maia ran off to the two huts, waving and still talking, leaving us to wonder what she had been telling us.

Our anchorage was such a perfect one that we hated to go. We sailed the dinghy, swam and did nothing, all very pleasantly. The wind died away in the afternoon and we motored round to Delos.

Everything was right for our visit to Delos. We had it to ourselves and the moon was full. Pietro brought the French Guide Bleu ashore with him, and we walked through the ruins in the still of the evening. We were amazed by their state of preservation and their vast extent. I had expected to see one large temple with some overgrown columns and walls fallen around it. Instead, we found walls and doorways, roofs and columns, floors and staircases, not of one temple but of dozens and of different gods and different religions. There were market places and meeting places—wide courtyards and deep cisterns. There were altars, statues and carvings, and designs in tiles and mosaic. Their beauty was made even greater because they were in the natural countryside among tall grasses, trees, bushes and wild flowers and not behind fences and iron bars. They had life in them, too, the wild life of lizards and frogs, hawks and singing larks.

We watched the sunset from the top of the mountain, near the Oracle's Cave. This place was quite unlike anything else at Delos. It was a natural cave, a jagged cleft in the mountain-side, with no pretensions to art or beauty, high up and far removed from the noise and bustle there must always have been in the ancient living Delos. We could imagine how a Greek, Egyptian or Phoenician might have grown weary or bored with the sophisticated religion of the smooth marble temples in this crowded holy place of pilgrims and merchants. Maybe at this rough and lonely altar, where the voice of the Oracle sounded through smoke in the hollow of the rock, he could feel himself closer to some divine power of nature.

While it was still light we walked down to the shore, sometimes choosing separate routes as the paths and passages twisted and diverged. The frogs were croaking and splashing into the cisterns as we passed. At one place I heard some different noises, not as loud and monotonous as the chorus of the frogs. I turned alone down an alleyway to find what it was. In a wide roofless courtyard were nine very small owls. Their flat-topped heads and round staring eyes looked much too big for their minute bodies and delicate legs. They were strutting about in twos and threes between the shadows of the pillars and the light of the marble floor, conversing in sweet little busy chattering notes. They walked nearly upright, leaning forward slightly, sometimes giving two or three quick little bows. In their fussy dignity they might have been a collection of elderly professors, their coat tails tucked up and their hands clasped behind their backs, discussing some heretical new approach to classical study.

We saw the light fade and then sat beside the columns of the market-place to see the moon rise. We rowed out to *Leiona* in silence. There was nothing to say.

The time for leaving the Cyclades was drawing near. In June the winds had been light and variable, and the nights calm. With July the wind began to blow more steadily from the north, and the blows were stronger and longer. We were told that we would find more comfortable sailing in the gulfs west of Sounion.

From Delos we sailed south and west, calling again at Paros and then going to the westerly islands. We called in at the volcanic bay of barren Milos, anchored among the oleanders of Sifnos, and stayed three days in the long inlet at Serifos.

The island's town, Castro Serifos, was on the top of the central hill, 1,500 feet high. Like Santorin, the way to it was up a mule track and steps, but it was a more gentle climb. The first half-mile was on the flat, through gardens of flowers and vegetables, along paths flanked by blackberry bushes, shaded by plane and olive trees, and overhung by branches of limes, apricots and mulberries. Our progress was slow and our fingers were stained by the purple mulberry juices.

At the foot of the hill we met the track winding up the bare rock, some-times as a pathway and sometimes cut into wide, long steps as a guard against being washed away. As the track twisted round spurs and over saddles, it gave us sudden new vistas of hill, valley and sea, good excuses to stop for admiration and rest. An hour of hot walking brought us to the town.

Castro Serifos was a striking village. Its houses were built on the sharp crags of the peak, close together, almost on top of each other, so that a woman scrubbing her white doorstep could be talking across the narrow street to a neighbour below her, hanging out washing between the chimneys on the flat roof of her house-top.

In front of the doors of a big church a man came and spoke to us in English. He was from the iron mines on the south of the island and had come to spend the day with his family to attend the wedding of a friend. We were taken into the family circle, first going from house to house, sitting in kitchens or front parlours, where we were given glasses of wine or liqueurs, and then to join a crowd in the courtyard of a taverna. The wedding was to be in the afternoon. We met the bridegroom who was already in his best dark clothes and was having his last party of bachelor-hood. We added our share to the piles of full and empty beer bottles that were gathering on the tables, joined in the laughter and listened to the songs.

Georgios, our friend from the church door, wanted us to stay for the wedding, but we meant to sail before sunset, and it was obvious that the wedding festivities would start late and go on well into the night. So we left the bridegroom to his friends and walked down to the harbour.

That evening we left the Cyclades. Ahead of us was a hundred miles of open sea until we came to the mouth of the Gulf of Argolis, with Nauplion at its head.

CHAPTER SEVENTEEN

South of Salamis

The flash of the Cyclops Point lighthouse on Serifos dropped astern in the dark. That night the wind did not die away, but came unexpectedly from the south, and we had a fast sail under a clear sky and a bright half-moon. I woke an hour before midnight and came on deck to see if Sandra was feeling tired.

She was sitting on the stern cabin, above and beside the wheel, her head high in the warm south wind and her eyes shining. The boat was swinging along with a little spray flying, and we seemed to be moving at a tremendous speed through the white crests of the black seas. Sandra was loving it. She said:

'This is wonderful. I've never known anything like it. Just the moon and myself to take the boat through these glorious waves. Don't stop me steering now. I want to go on and on.'

I left her there alone to finish her watch, and then she was glad to have a rest and let Paolo take over.

We stopped at Spetsai and then spent the rest of July in the Gulf of Argolis. I particularly wanted to see Spetsai, as a possible place for spending the winter. This problem was always on my mind and I had not seen in the Aegean a place which was safe, convenient and pretty. In the Ionian Islands I had marked Argostoli, Gouvia, and even Gayo, as possibilities, but I really wanted to be on the eastern side of Greece for the winter.

I was already thinking about the plans for the next year, and Rhodes and Istanbul were attractive targets, if I could fit them both into one summer. A whole summer sounds a long time, but in the Aegean a lot of sailing time can be lost during the Meltemi season in July and August. I would get my best sailing in the early and late parts of the summer, and so it would help if I could find a winter base near the cruising area.

Spetsai had all the qualities I wanted. The wooded island was pretty, and the town was big enough to provide shopping and facilities, but not to be an overwhelming city. The harbour at the town did not attract me, but there was a well-sheltered inlet close to it. In this inlet were slipways and building yards for local caiques, and they could deal with *Leiona*.

It was comforting to have one good wintering place marked down, but I still had three months of sailing to come before the winter, and there were many other possibilities to explore.

For two weeks we wandered leisurely around the big gulf, anchoring off beaches, in empty coves, or in village harbours.

Paolo was in a hurry to get to Nauplion. He thought it was high time we had some really good sophisticated meals ashore. As the full-time, hard-worked cook he felt more strongly about this than the rest of us. We had tried for luxury at Spetsai but had failed, probably through not knowing the right places to go.

Our last effort had been a great disappointment. While looking for somewhere to dine we had first smelled and then seen an appetizing round of meat being grilled on a spit at an outdoor restaurant. It looked like a roll of veal encased in crackling as if it were a sucking pig. We took a table beside it and waited to be served.

The sight and the smell and the sound made us impatient to start eating. At last a sizzling length was put before us on a big platter. We sliced through the golden crackling and uncovered a tangle of white and tasteless 'tubes', apparently chopped up veins and intestines. Costanza's face took on an expression of horror and disgust, and we all felt the same. We made some half-hearted stabs with our forks and nibbled at the crisp outer coat and some of the less tube-like inner parts, but we had lost our enthusiasm.

Greek food is not to everyone's taste. We had only been visiting
8

villages and small towns, and so had no experience of top-class menus. In Argostoli, our first Greek town, we had found a good simple restaurant, and also in the Aegean at Mykonos and Naxos, inspired by tourist trade. All these had given us reasonable meals at reasonable prices. It was very different in the villages. There the price was low, but so was the quality. To get a good meal was an interesting exercise, involving a good deal of effort.

The village tavernas did not have menus, which in any case would have been little use to us. The custom was to go into the kitchen and choose from the food being cooked there. This was only the first part of the effort. When we were new to the problem we used to select our whole dinner at the one kitchen visit. This was a mistake. After a long wait all our selections arrived at once. We had seen the food hot and bubbling in the saucepans, but by the time we got it most of it was cold. The little that was still warm had time to cool while we tackled the first course.

I had forgotten the advice which Bobby Somerset had given me about how to avoid the worst in Greek food. There were three points to remember. The first was to order only those dishes which were to be eaten straight away. The second point was to remember the word *zestos*, and to use it frequently and from the start. *Zestos* means 'hot', and to get food served hot is not easy in Greece. We would choose it while it was hot, and still on the fire, but unless we kept on convincing the waiter that we wanted the course served hot, and served at that moment, he would keep it until it had cooled down to what seemed to him a reasonable temperature.

As the food had been cooked in a lot of olive oil, it could become greasy and cloying as it cooled. This brought us to the third point, to use the phrase *ochi ladhi*, meaning 'without oil'. This was not to avoid the use of oil entirely, as it was the only medium for cooking, and a good one, but to stop the waiter pouring a jug of cold oil, like gravy, over a plate of food as he served it, cold or hot.

Some of the food was good in the villages. There was not much variety, but we usually found grilled lamb chops or roast chicken, meat on a spit, tomatoes stuffed with savoury minced meat and onions, vine leaves wrapped around rice or meat, and *moussaka*, a kind of shepherd's pie with aubergines. There was always fish, and usually squid or octopus, grilled, fried or boiled. Fruit and vegetables were good, the choice changing with the seasons. Bread differed in each island, and was usually good. The only staple food we had to buy which we did not like was the cheese, which went hard and dry very quickly.

The common wine in the villages, and sometimes the only kind to be found, was *retsina*, a white wine with a strong taste of turpentine. Only Pietro and I liked it, especially as a well-iced midday drink, but I found it a pleasant change to drink ordinary wine again. The normal 'quick drink' in the villages was *ouzo*, a transparent spirit which turned cloudy when water was added.

Later in the year, in the northern islands, I found the food and wine better, but this may have been because I learned what to look for, or was getting used to it.

Paolo was right. Nauplion did provide some really good food, in restaurants both in the city and in the hotel on the island-fort of Bourdzi at the mouth of the harbour.

But we remembered our stay at Nauplion for better things than just food. My crew went off for a day at Mycenae, and came back full of praise and pleasure. I had not felt like leaving *Leiona* untended for so long, as some gusty winds were blowing and I did not know how good the holding was in the soft mud of the harbour.

Dominating Nauplion was an impressive Venetian fortress on the big hill above the city, looking out to sea and over the harbour. I could explore it, and at the same time keep an eye on *Leiona* and watch for any change in the weather. So I set off alone for the citadel. It was a steep climb, up a stairway which I was told had 999 steps. I forgot to count them, but even when I was only half way to the top I was ready to agree to the mystic number, and maybe add to it.

The views from the top were breath-taking, and the fort itself was fascinating. It was a labyrinth formed of double and triple walls with tunnels and corridors joining a complex pattern of strongpoints, court-yards and gateways. On the seaward side the cliffs were often vertical or overhanging, and I could look straight down into the shallows and rock pools below. The water was so clear that it was hard to tell which of the fishing boats were hauled up on the strip of beach and which were floating.

August took us into the Saronic Gulf, the big island-dotted square of water with its four corners resting on Piraeus, Sounion, Poros, and Corinth. As always, we wished that we had more time. In two weeks my Sicilians had to leave and go back to their farms and businesses and families. We found the two weeks not nearly enough for all the variety and pleasure our last cruising area produced for us.

On our way to it we called at Hydra. I felt that I had been there before, because it was so like a miniature Ischia. It was a perfect romantic picture of a Mediterranean harbour. There was a narrow entrance between a steep headland and the end of the sea wall. Then we found ourselves in the town square, but it was a square made of water. Three sides were enclosed by gaily coloured shops, restaurants and houses, seeming to grow out of each other's rooftops on the steep hillside. Only a pavement and a road were between the houses and the waterfront, which was crowded with local boats and caiques, all brightly painted.

Hydra was pretty and gay, but one night there was enough for us, and we sailed on past the long single ridge of the island to round into the Saronic Gulf.

Our last two weeks together were very happy and carefree. We were within easy reach of Athens when the time should come for my crew to pack up and leave me on their return to Sicily. We were not tied to any programme, and we moved to suit our wishes and the day's wind. The days were hot and the winds were light, so we spent our time in idle sailing, swimming and fishing. Each day was pleasantly like the other, and all our evening anchorages were delightful. The whole of this time was a lazy, hazy dream, and only Poros, Epidaurus and Aegina stand out as clear memories.

'The sheltered channel between the mainland and the green island of Poros had many pretty bays for our midday swimming and fishing. When the two girls were not trying to increase the tan of their already cigar-coloured skins they used to join Paolo and Pietro in their under-water hunts. I tried to catch up with letter-writing, but my far-away thoughts would be interrupted and surprised by what looked like a school of sea monsters. These were the four swimmers surfacing together, breaking the water with splashing flippers and spouting breather tubes.

Letter-writing would be left while I watched them swimming in the clear water below and around the boat, flashing and shining as they moved and twisted between shadow and sunlight. The girls looked like mermaids as their long hair trailed behind them. Paolo added touches of mythology, surfacing like Poseidon with his trident, and sometimes raising on the trident an octopus looking like Medusa's head with writhing snakes as its tentacles fought to get free.

Paolo had wondered where we would meet our third 'threat of war'. Poros was the place. On one moonlit night, after supper ashore, we walked and clambered up the twisting alleys and footpaths of the old part of the town, to get to the top of the hill. We wanted to see the silver of the moon on the olive groves and on the open bay facing the Aegean. The houses stopped just short of the top of the hill which was rounded and rocky, covered with a few bushes and small trees, and with only one building, a white tower. It looked like an old mill, without any vanes.

On our way to the mill we were checked by a wire fence, but it was so old and broken that we got across it easily. Near our crossing place was a small sign board on a pole. We tried to see what it said, but the writing was away from the moon and we could not read it in the deep shadow.

We stood near the tower to look at the beauty of the silver olive trees and the shining sea. Then we started back, crossing the fence at the same place. By then the moon had moved round enough to shine on the sign board. There were two Greek words. I knew that the top word meant 'Attention', but I did not understand the lower one. We thought it was probably something like 'Private Property', but it was too late to obey it. We were wrong. I found the word in my dictionary, and it meant 'Minefield'.

The memories of Aegina and Epidaurus were less frightening.

At Epidaurus we anchored in the little inner harbour, too shallow for most boats. It was a pretty and enclosed harbour, with a small village at the mouth of a valley below a steep mountain. We had come to see the old amphitheatre, an hour's drive inland, and we found that we had come on a day when there was to be a concert in the theatre, given by a German orchestra and choir.

During the afternoon our quiet little harbour became busy with launches from large yachts anchoring in the outer bay, and with relays of small steamers bringing in the audience.

We decided to join the audience, and we had a hair-raising taxi ride to the theatre. The driver went as fast as he could, and on the mountain road it was too fast for us. We had become used to the slow progress of a boat. The normal dangers of the winding road and steep hillsides were increased by the concert traffic, and by the frequent roadside shrines. Our Dimitrios was a devout man and he crossed himself with concentration as we passed each shrine. Maybe he felt the need for divine protection. Unfortunately, most of the shrines were at prominent places at sharp bends and above the steepest slopes. It was frightening to meet a returning taxi, also going fast, just as Dimitrios let go the wheel and bowed his head. It was a relief when the journey was over.

The Sanctuary of Epidaurus was crowded when we arrived. Maybe that was the proper way to see it, because that was how it must have been when it was a living place over two thousand years ago and people flocked there to be healed by the powers of Apollo's son Aesculapius.

We walked about in the gardens and among the ruins, and at sunset we joined in the flow to the theatre. At the foot of the theatre we looked up at the slow-moving mass of fifteen thousand people spreading upward into the huge bowl. The widening half-circle of stone seats were floodlit, and were in bright contrast to the blackness of the hillside behind them, where the sharp peaks stood out against the fading colour of the sky.

Our seats were nearly at the top, yet we could hear the music clearly, without any need for amplifiers and loudspeakers. The audience filled the theatre, and merged in with it. The only discordant note was the appearance of the modern evening dress of the orchestra and choir, and I wished that we had been looking at a production of ancient Greek drama in classical dress.

Aegina produced unexpected pleasures. We had gone there for fresh food and did not expect to like so 'suburban' an island, with a service of fast and frequent steamers to and from Piraeus. In fact, we enjoyed its town, its coast, and its countryside. The walled harbour was small and shallow, but it was good enough for *Leiona*, and we found a quiet place on the sea wall.

The steamers kept to the entrance end of the harbour, and the rest of it was taken up by fishing and market boats. The market boats all gathered in one part of the quay, opposite the shops and restaurants. They came into the harbour laden with fresh fruit and vegetables, and sold their loads either in bulk to the shopkeepers or over their scales to the housewives. Their part of the quay was a crowded place, and it was full of animation, sweet scent, and colour.

The most impressive sight in Aegina was the beautiful temple of Aphaia, above its eastern coast. This was a magnificent temple in a splendid setting, looking right across the Saronic Gulf to the equally magnificent temple of Poseidon at Sounion.

Aegina's bays and beaches were our last quiet and secluded anchorages before we went back to the bustle of Piraeus. In Turkolimano we were lucky to find a place, apparently the only place, for *Leiona* at the jetty. We dined very late that night, our last night together, at a table by the water's edge. The surface of the harbour was still and as black as polished ebony, reflecting the lights from the encircling sides and the stars in the opening of the harbour mouth.

The next day *Leiona* was an empty boat, and I was alone for my sail to the northern islands of the Aegean.

North to the Journey's End

I had two good reasons for going north. I wanted to visit new waters in the two months of summer still left, and I wanted to see Michael Osborn. Michael was a close friend whom I had not seen for five years, and he was not far away. He was living on the island of Skiathos in the Northern Sporades, and for the last year we had been writing to arrange a meeting. He had added a third reason for my northern journey, suggesting Skiathos as a wintering place, and offering his house as a shore base for me.

I had still not decided on a place for my winter home. My list of possibilities had increased, but without yet including a certainty. So far Spetsai still seemed the most practical choice. Epidaurus had a romantic attraction, but it was too lacking in facilities. I had liked both Aegina and Poros, although neither had a slipway suitable, for which I would have to go over to Piraeus. And Aegina's harbour looked as if it would be bleak and uncomfortable in the winter gales. There was shelter in the Poros channel, but the only convenient anchorage was in the Naval Academy waters, and I was told that the commandant no longer welcomed foreign yachts.

Obvious places with plenty of facilities were Turkolimano and Passalimani, and the new harbour developed specially for yachts at Vougliameni, ten miles to the south of Piraeus. If I had been leaving *Leiona* under paid supervision for the winter I might have chosen one of them, but I did not fancy them as places in which to make my home.

As well as Michael Osborn's suggestion of Skiathos I had been told of two possibilities in the area I was planning to visit, the island of Kea and the town of Chalkis. No one had recommended Volos, but from the chart the enclosed Gulf of Volos, not far from Skiathos, looked a possibility, with the city at its head and several apparently sheltered small bays within the big gulf.

The only thing to do was to have a look at all of them.

My plan was to visit Kea first, and then to go on north. Trying to move northward in the Aegean during the Meltemi season, which still had a month or so before easing, did not seem a good idea. It would mean working against an almost constant head wind and disturbed seas. Luckily, there was a sheltered route for my September journey.

Nearly all the eastern shore of the Greek Attic mainland is sheltered from the north winds of the Meltemi by a long ridge of mountains. This ridge runs the whole length of the narrow island of Euboea which lies close along the mainland coast. Between the island and the mainland runs a channel nearly a hundred miles long and seldom more than four or five miles wide, with Chalkis at its halfway point. This channel would take me in sheltered water most of the way from Sounion to Skiathos. Even if there were some parts which the wind could reach, there would not be room for any big seas to develop, and I could find plenty of anchorages.

There was no urgency. If I hurried I could reach Skiathos in three or four days, but I could see no advantage in arriving at the exposed Sporades while the strong Meltemi winds could still be expected.

It was hot in Piraeus in August, but I could find at least cool evenings and nights in the bays off Salamis and down the coast between Piraeus and Vouliagmeni. I had enough to do to keep me busy around Athens until September. My occupations stretched from the sublime of classical sights of the Acropolis and the Parthenon and the museums to the ridiculous, or at least more domestic, of changing oils, topping up fuel and water, loading liquor from bond, and having gas bottles changed or filled.

By September I was impatient to get going, and I sailed off on my own. I spent the whole of the month on my way to Skiathos. It was a month of gentle sailing and pretty anchorages. I enjoyed the quiet of sailing alone, although I felt guilty and selfish in not sharing the pleasures.

Kea was better than the *Pilot* had led me to expect, with its description of it as a coaling harbour. In my mind I had a vision of it which must have come from childhood memories of reading *The Wonder Book of Ships*. I could see very clearly a cluster of grimy ships in a haze of coal dust, with lines of blackened men trotting up gangways, carrying baskets of coal on their heads.

There was nothing like that in Kea. Coaling activities had stopped many years before, and although the coaling sheds and docks were still there, they were silent and deserted at one corner of the harbour and did not inflict their presence on the view.

Two headlands nearly joined at the entrance, and inside them the harbour spread out to either side within a ring of hills, forming two separate bays. The anchorage in the south was more convenient, near a large village, but the northern bay, where there were only a few houses, gave better shelter. I anchored in the north and had a pleasant walk to the village along the road around the harbour.

Kea provided a safe harbour, but there was no slipway, and for any technical shopping or help I would have to go to Piraeus, either in *Leiona* or by caique to the mainland and then by train or bus. I liked the island and the harbour, but I thought it was not really practical enough, and I hoped I would find something better.

September in the shelter of Euboea was a very pleasant month. My solitary wanderings took me to places with poetic names like Amarinthos, Boufalo, and Artemision, and with the great historic names of Marathon and Thermopylae.

At Thermopylae I walked to the statue of Leonidas on the site of the old battlefield of The Three Hundred. It was on a plain below a steep slope, and I did not find what I had expected, a narrow passage where the few had checked the Persian army. Later I saw a historical map which showed how in the last two thousand years the sea had receded from the foot of the mountain against which it had touched in the time of Leonidas.

I carried out my only rescue at sea on the day I left Kea. It was not a very exciting operation, but it brought me to a pretty harbour. It took place on a hot and glassy afternoon, following two days of strong north wind, when I had lain comfortably in Kea's northern bay watching the big seas march past the harbour mouth.

On a morning of light wind I started for the Petali Islands, at the southern end of Euboea. The wind died away and by the afternoon I was motoring. For some time I had seen a white sail ahead of me, but nearer to the mainland. I thought it had been luckier in finding more wind than I had. After a while I could see that it was not moving at all. It had been in the same place for a long time, and I closed towards the boat to see if any help was wanted.

The boat was a little caique converted to a yacht, and in it were three very nice Athenians who had left Ormos Raftis for a morning of fishing and sailing. By noon the wind had left them and they started to motor home. Then the engine stopped with a broken fuel pump. They had spent several hot and frustrating hours with no wind, engine, or even oars, and no other boat had passed within waving distance. They were delighted to have a tow back to Raftis.

I was glad to have had this little adventure, as otherwise I might not have gone to Raftis, which was a charming village in a pleasant bay. In the middle of the entrance stood a big rock surmounted by a pillar on which was seated a statue without a head. I was told it was the statue of a tailor, which is the meaning of the bay's name, but I could see no indication of his trade, except that the figure was sitting with crossed legs.

We exchanged visits between the two boats, and then I rowed the Athenians ashore, to wait under the mulberry trees of the village taverna until a car came to collect them.

They were the only people I met to speak to until I reached Chalkis a week later, after calling in at the Petali Islands and the small bays on the Euboean shore. Chalkis was a possibility for a wintering place. It was a pleasant town, with easy road and rail access to Athens. *Leiona* would have been able to lie at anchor in shelter south of the town, and I could find a slipway nearby. I marked it down as a place to remember if Skiathos or Volos did not turn out to be suitable.

I knew that there was a bridge at Chalkis which only opened for a short time at each six-hourly turn of the fast local tide, but I could not find in my books and tables the timings of the tides. I wanted to pass through to the northern side as soon as I could so that I would be able to leave whenever I felt like it. I was resigned to a wait of a few hours for the bridge.

The bridge was closed when I arrived, and one yacht and two schooners were anchored on the southern side. I put the dinghy overboard and went to the port office to pay my fee, a very small one compared with what had been charged in the Corinth Canal. I asked at what time the bridge would open. The sailor at the desk gave me a rather despairing look and said, 'It is ready now.'

It must have opened while I was walking from the dinghy to the office. I hurried back to *Leiona* and got the anchor up in a rush. I hate towing the dinghy in a narrow place of swift waters, especially when I am alone, so I dumped it unlashed on the cabin-top. All this was done quickly, but even so the yacht and the schooners had disappeared, and another schooner was coming through from the north. The bridge was starting to close. I kept on going towards the bridge, and used for the first time in earnest a Spanish hunting horn bought in Palma. A sailor in charge of bridge operations looked doubtfully at me and then opened the bridge just in time.

I went on with my wanderings in the North Euboean Gulf, which forms the greater part of the channel. All the channel was beautiful, between the steep slopes and close mountains of Euboea and the gentle slopes and bays of the Attic coast and its distant peaks. Every evening was calm with striking sunsets and the nights were warm and clear. By day there was always something of interest to see, with villages appearing on either shore, and with fishing boats, schooners, and an occasional steamer moving along the channel.

I had begun to lose count of days and dates until I realized that October had come, and it was time to move on to Skiathos.

I passed between the north end of Euboea and the entrance to the Gulf of Volos, and then I was in the open sea, with Skiathos only a few miles ahead.

It was exciting to see Skiathos. I was coming to a place about which I had been thinking for many months, and which might be my home for the next winter.

Daylight was ending by the time I reached the island. I did not want to try finding my way into a new and small harbour in the dark, so soon after sunset I anchored in a wide bay on the southern coast, off a white beach below a pine wood.

Skiathos was the prettiest village I had seen in the Mediterranean, and I came to it on a perfect October morning. The view up the long bay opened in sudden beauty as I rounded Cape Kalamaki and began to tack to the north. To port was a line of coves and cliffs below the green and wooded ridge which formed the long arm of the bay. To seaward a chain of islands ran north to join the short arm of the bay. Two miles ahead of me the arms joined to make a narrow pocket where I could see a few houses and some boats hauled up on a beach. West of this pocket was the village of Skiathos, white houses and red roofs among the big green trees of three adjoining hills.

As I tacked up the bay, working close to the longer shore, I saw scattered houses built on the slopes above the coves, and I wondered if Michael's house was one of them. A white-sailed dinghy was launched off one of the beaches and sailed out into the bay. After a while we crossed tacks and I could see a man and a woman sitting out to windward. They passed near me and a very English voice called out:

'Are you Keith Robinson? I am June Snowball.'

She was someone I had heard about for many years without ever meeting, a friend and neighbour of my sister in England.

This was a pleasant welcome to Skiathos, but it was also something of a shock. I had forgotten how much I had become separated from my life in the army and my English friends. During the last six years I had only spent six weeks in England, and since owning *Leiona* my life had become concentrated on her, in summer sailing and winter preparation. The connections and thoughts evoked by the words of a stranger took my mind back in a giant's stride over a gap of three years and more.

I was not sure where to anchor in Skiathos. There were two harbours, or anchorages, in front of the village, separated by a wooded hillock at the end of a narrow neck of land. I could see no yachts anywhere, but there were a dozen caiques in each harbour, anchored just off-shore in one, and lying stern-to in the other, against what seemed to be the main street. I looked into both and decided on the open anchorage, north of the hillock.

Michael and Anita Osborn found me as I walked around the village wondering how to get in touch with them. I had got as far as discovering that there were no telephones and only one taxi, which was busy some-where else.

I had already been enchanted by the village, from the very first words spoken to me ashore. I had rowed in to the quay, taking two water cans for filling. Two young boys had taken the dinghy's painter. I asked them where I could find water and received the delightful reply:

'The spring is under the first plane tree. But we will fill your bottles for you.'

My guess about Michael's house had been right. He and Anita had had one built to their own design, and were now spending more time in Skiathos than in England.

They adopted me in Skiathos as Paolo and Costanza had done in Palermo, and I was swept up into the social whirl of the island, this time in an English society. It was fun to meet a lot of happy and friendly people, and life seemed strangely easy, almost too easy, without any problems of foreign languages.

This little English colony lived three miles from the village, so I moved *Leiona* to the cove below Michael's house, a very pretty anchorage, but open to the north-east and with a weedy bottom. Twice I had to move out in a hurry when an early morning breeze set up a steep sea, and I was too close to a lee shore to accept the risk of dragging.

It was all lots of fun, but as in Palermo the summer before, social life began to pall after a while, and I wanted to get sailing again. I tried to take Michael off for a week or two, but he was involved in so many activities that he could never be free for more than a day at a time. Ted and June Snowball could come, and we had a delightful few days in the Sporades.

It was just what I like doing best of all, sailing with two friends among small islands with no programme and no pressing dates. We had wonderful weather and pleasant short sails between pretty and peaceful anchorages. We enjoyed ourselves so much in the group of islands stretching eastward from Skiathos that we did not get as far as we had planned. I had wanted to go to Skiros before my summer sailing stopped, but the attractions of Skopelos and Alonnisos proved too strong, and we shrank from the 'long sea voyage' of the twenty-five miles to Skiros.

When we came back to Skiathos I had to think seriously about my winter plans. Luckily three days of strong wind came along which gave me a picture of winter conditions and helped me in my decision.

I liked Skiathos better than anywhere else I had been in the Aegean, and my first impressions were like those I had about the village and people of Puerto de Andraitx. It was pretty and sympathetic, and it had a slipway big enough for *Leiona*. But I was not happy about the convenience and safety I would find there.

The safest place, and maybe the only practical one, was half a mile from the village, across the pocket at the north end of the bay. Even there I was doubtful about complete security, and I did not like the idea of winter dinghy trips in the darkness, rain, or wind.

Safety was the most important requirement, so I left Skiathos and sailed for Volos, at the head of the big Gulf of Volos. On the way I stopped for one night in a small cove which I had noticed on the chart, and liked, many months earlier. It was in the south-east corner of the gulf, sheltered by a little island across its mouth. It seemed a perfectly safe anchorage under any conditions, and I was pleased to see a good slipway there, even though the village had only a few houses. I felt that if I did not like Volos as a place to stay, I could use this cove where *Leiona* was sure to be safe, going to Volos by land or sea if ever I needed something the village could not supply.

I did not expect to like Volos. It was a large town, almost a city, and it was a steamer port. I thought it would be crowded, noisy, and dirty, with oily water. *Leiona* would have to lie with her stern made fast to a main street, and with memories of Ischia, the only other place I had tried such a mooring, I thought a winter in Volos would be unbearable.

I was wrong. Volos was an amazingly good place for wintering. It fulfilled all my requirements except that it was not pretty. An earthquake had flattened it ten years before, and the hurriedly rebuilt town did not have charm or beauty.

It was not entirely without attractions. Mount Pelion was close above the town, and its steep face was livened by the differing colours of strips of woodland, watercourses, valleys, grey rocky screes, villages, and the climbing red snake of a new road. On the opposite side of the harbour a wooded hill and headland held the ruins and old walls of ancient Pagasae, from which the Gulf of Volos has its Greek name of Kolpos Pagasitikos.

Beyond Pagasae was a line of higher hills fading away to blend with the northern mountains of Euboea, across the channel. A walk up the lower slopes of Pelion brought the whole gulf and its almost hidden entrance into view.

Sunsets spread soft lights of pink and mauve over the hills and the quiet waters of the harbour, and as the sky darkened it was hard to tell the difference between the stars and the window lamps of the villages on the heights of Mount Pelion.

From the more material point of view Volos was admirable. Within the headland of Pagasae and half a mile of new breakwater a big harbour had been dredged out of the shallow head waters. This harbour was divided by a long steamer pier which jutted into the middle of the deepened area, carrying cranes, a huge warehouse, and a railway track.

West of this pier was the cargo port for local caiques. It was a crowded, busy and fascinating port. Most of the boats were from forty to sixty feet long, in a variety of shapes and colours. Red, white, and orange were the favourite colours, picked out with lines and patterns of blue and green and brown. A few of the boats were schooner rigged with two masts, but most had only a foremast, used for their cargo derricks and for setting a jib or staysail to steady the boat, or as a desperate auxiliary if the engine should fail.

All these caiques were crowded into the port, some stern-to the quay and some alongside in single or double banks. There was a splendid old-fashioned Elizabethan look about this gathering of rounded wooden hulls, helped by the thick shrouds with their deadeyes, lanyards and ratlines, and by the big anchors hanging from the catheads.

There were transom sterns, pointed sterns, and short or long counters. There were bowsprits and bald stems, and straight bows, curved bows, or clipper bows.

The cargoes of goods, animals, or people were just as varied as the shapes, and so were the forms of transport which loaded the caiques. Shining big Mercedes lorries would be standing side by side with pack mules, gay pony traps, or narrow unpainted four-wheeled carts drawn by single horses whose bells and coloured necklaces jingled as they shook their nosebags.

A few trawlers shared the cargo port with the caiques, but the twenty or more net-fishing boats lay on the other side of the steamer pier, along the Street of the Argonauts. I found a place for *Leiona* here, separated by four fishing boats from the only other yacht in the harbour. This was *Coppra*, a 48-foot ketch, American-owned, but built and registered in England.

The Street of the Argonauts was half a mile long, and it was wide and clear, in front of a line of white buildings, nearly all cafés, restaurants, or hotels, except where the Ionic columns of the Bank of Greece stood out in pale brown contrast. The street was free from noise or traffic. It had a wide roadway and two wide pavements. The landward pavement was filled by the tables and chairs of the cafés, and the pavement beside the sea was almost always covered by the long dark brown, yellow, pink or orange nets, laid out for their daily repair. With no pavements to walk on, the population had to promenade along the roadway, and so no cars were allowed on the street.

I am finishing this last chapter beside the Street of the Argonauts, on a January afternoon, on Twelfth Night. It is a good winter. Margaret and Oury Hisey, the American couple living on board *Coppra*, are good neighbours. Our problems and our way of living are very much the same. In spite of this we usually only meet by chance two or three times a week, and so we are glad to see each other.

The convenience of living in Volos for the winter has been greater than I could have imagined. The shops I need, the post office, and the bank are all within five minutes' walk from my stern plank. There is a water pipe two yards from me, and even main electricity and telephone, if I wanted them. Across the harbour is a slipway which the owner assures me can take *Leiona*. It looks too shallow and small to me, but we shall see in the spring. There is always the other slip in the cove at the south of the gulf, twenty miles away.

The town is too big to give me the clear and concentrated pictures of its personalities that Andraitx did, but I have found the people kind and helpful. The Harbour Master and Pilot came to see me and pressed on me the use of a large mooring buoy nearby. Now with two anchors out and a double line to the buoy *Leiona* should be safe from the winter gales, which have not yet come.

It is odd how patterns of life repeat themselves in different places. In Volos a dynamic young Greek manufacturer and his family have taken charge of me and given me help and advice as Bobby Somerset did in Andraitx, Paolo in Palermo, and Michael Osborn in Skiathos. The cast of the new play has even provided me with successors to Bartolomeo and Josefa.

For the last two months I have been on waving and speaking terms with an elderly fisherman who comes and goes at intervals of several days in his little white cabin motor-boat, sometimes alone and sometimes with his wife. Last week, without any suggestion from me, he surprisingly announced, as an accepted fact:

'In the spring my wife and I will come and help you clean and paint your boat inside and out. It is a good boat and we will take good care of her. And it will be for little money.'

So I shall have my 'married couple crew' again in Spiros and Artemis.

A year ago in Andraitx I was watching for the arrival of the Three Kings on this Twelfth Night evening. They have been disregarded here, and instead there has been the Blessing of the Waters. It has not been the most fitting day for the ceremony. For the first time the slopes of Pelion are covered with snow, and since early morning a cold north wind has been blowing.

Probably because I lent some flags to help the Harbour Master decorate the ceremonial barge, I found myself on board with the priests, prominent citizens, officials, choir, and incense bearers. The cross had been escorted down the crowded street by a dignified and richly-coloured procession as far as the front of the Bank. Then some of the escort bore the cross on to the barge, and we were towed away from the shore. While the chanting was growing stronger I saw some of the sailors in the towing launch taking off their overcoats. The cold wind was still blowing, and flakes of snow were melting into the water.

The flowered cross was then held high over the water, and the shivering sailors stripped for swimming and stood at the gunwale. Then the cross was thrown far out from the boat and the swimmers dived in and raced for it.

I hoped that *Leiona* had been included in the blessing. My few years of sailing and living in her have made me sure that this is the kind of life I want. I would not now like to change it for any other. The past years have been good. Now I am thinking of the years to come. Rhodes and Istanbul and the Dodecanese would be good targets for next summer.

And then perhaps Venice and Dalmatia after that, and back to Spain for the winter. Or a visit to the Caribbean. Or a summer in the French canals. Or the Balearics again. The scope is so wide that it is hard to choose. Maybe I will try them all.

My new life has had a good beginning.

APPENDIX

Prices and Problems
in Mediterranean Sailing

I first wrote this appendix in 1964, and at that time I wrote it in great
confidence and detail, meaning it to be an accurate guide to the expenses
of living in a boat in the Mediterranean. I am re-writing it in 1967, in
Mallorca again for a while after two years in Greece, and now I see what
great differences are made by changes of place, time, and conditions. An
attempt to achieve accuracy would only be misleading. I hope that the
very general figures and ideas, and the few personal examples I give will
help the reader to form an estimate of his own likely expenses.

The one sadly unchanging fact is that all expenses are increasing, and
the increases are rapid.

PRICES

BOAT EXPENSES

My boat expenses fall most easily under the headings of Insurance,
Slipping, Labour and Materials. They are for my 17-ton cutter, and would
increase or decrease with size of boat and differences in local conditions.
The figures for labour are debatable, because they could be reduced to
nothing if I did all the work myself, but once a year, for a few days each
spring, I find it worth paying for some help.

Insurance

This costs me about £100 a year, for twelve months in commission. I could reduce it by insuring for only six or eight months' sailing and the rest laid up, and in fact I did so for two years. The expected saving was only about £30, and each winter I found that I wanted to move the boat earlier than I had meant, involving subsidiary insurance which almost cancelled the saving. Now I live in *Leiona* for all the year and sail in the winter as well; the additional insurance is far less than the cost of living ashore for those few months.

Slipping

It is hard to find a constant figure.

I used to slip twice a year, but now I have changed to the more normal Mediterranean yachting practice of slipping only in the spring. I hope that I will not find it a false economy; I am ever conscious of the danger of teredo worm in a wooden boat.

I have not slipped in Italy, where the costs are said to be high. In Spain my slipping at a fisherman's slip in 1962 cost me £12 each time. Later the slip became unsafe and I had to go elsewhere. One extortionate yard charged me £42, and another much more efficient one charged £18. In each case this covered the hauling out and launching, with the labour involved, and up to a week on the slip, when I provided my own labour for work on the boat while hauled out. I have been quoted £30 for my 1968 slipping, but I may find a cheaper place.

For the same arrangement in Greece in 1965 and 1966 I paid £18 and £22, at a very simple but good slipway, on a beach where there was a 'village' of only one house. At the more sophisticated Greek shipyards around Piraeus or at Spetsai I probably would have had to pay about £40 or £50.

Labour

A big saving of expenses is to employ no outside labour at all, or at least no unskilled labour. Occasionally some skilled mechanical or electrical labour is needed, and this is expensive, but probably cheaper in the long run than an amateur do-it-yourself effort.

I have tried doing without help for rough work while slipped, but I found the time and effort involved was enormous, working alone. Now I always try to get hold of two local fishermen for a week or ten days each spring, for scraping, cleaning and painting the bottom and topsides while hauled out, and for work on the mast and rigging when back in the water.

I have found it better to have two men than one, because they can help each other in moving the 'scaffolding', usually some planks, barrels, and boxes, and handing each other pots and brushes without disturbing me. Then I can work on the more detailed jobs of engine, seacocks and steering gear.

The cost of labour has increased enormously recently, and I suppose it will go on rising. In Spain in 1963 and 1964 a fisherman would work for me for £1 5s. od. to £1 10s. od. a day. In 1967 the basic daily wage is £2 10s. od. or more.

In Greece my painters were pleased with £1 each in 1966, but in 1967 the lowest wage was £1 10s. od. a day. This was at my village slipway, and in Piraeus the rate would have been nearly double.

MATERIALS

Paints. Paint, varnish and anti-fouling probably make up the biggest recurring item of maintenance gear, and of course the expense varies in proportion to size of boat, and in the quality bought, although cheap paint is not always an economy.

Local Greek paints are not very good. It is easy to buy foreign paints, but duty and transport make them double their original price. I think it is worth paying double, or importing a duty-free shipment, with other gear, if the order is big enough to cover transport and agency costs.

Spanish paint, especially anti-fouling, is good, at the same price as paint in England.

Leiona is afloat and sailing all the year, and I try to keep her well painted, which involves a good deal of touching up or repainting in the late summer and winter. My paints, including anti-fouling, cost me about £50 a year, which should cover the expenses of oils, brushes, sandpaper and fillers.

Sails and Rigging. With terylene sails and cordage there seems to be virtually no yearly replacement. I have just bought a new suit of sails, although only one old sail was showing bad wear, and that after seven years of hard work in the Caribbean and Mediterranean. No cordage has had to be replaced. I re-rigged *Leiona* soon after buying her, and all the galvanized rigging looks good. The only replacements I make, every second year for safety, are the ¼″ diameter flexible galvanized cables of the main halyard and steering gear. The wire is cheap and the splicing is easy.

Hardware. I can't really account for them, but I always seem to be buying a few more screws, hooks, bolts, shackles, or lengths of plastic tubing. I think that my bill, including the flexible wire rope, would easily be covered by £25.

Fuel and Oils. During the last few years I have travelled over 4,000 miles each year, 3,000 of them under sail and 1,000 with the engine. So a fairly constant figure of 200 gallons of diesel fuel is a steady item. In Greece fuel is 2/– a gallon, but oils and grease are expensive. In Spain fuel is about 6/– a gallon, with oils at the same price as in Greece. In my Mediterranean wanderings a fuel bill, to cover oil changes, harbour running, and battery charging has averaged about £100 a year.

Summary. In very general figures, at 1967 rates, I think that *Leiona* is costing me, in cash paid out each year:

Insurance	£100
Slipping and Labour	£100
Paints and Gear	£100
Fuel and Oils	£100
	£400

And I think that another £100 should be allowed for irregular expenses which must come up every few years, such as new sails, a new dinghy, re-rigging, or engine overhauls. So that makes a total of £500 a year.

But that is not the end of it.

To give a true picture of expenses, as if for a business balance sheet, two more figures should be added, depreciation and the loss of income on capital laid down. Neither is easy to assess.

I don't think that *Leiona* has decreased in value, and in fact I have three times been offered for her a good deal more than I paid, but the cost of 'the next boat' will also increase. So maybe another £200 or £300 should be added to her yearly cost to balance the account.

The question of capital is still more tricky. *Leiona* and her gear cost me about £6,000, and I have since added more equipment. But if I did not live in her, I would have to buy or rent somewhere else to live, and would be tempted to own a car. So maybe it would be wrong to add a further potential expense figure of some £300 a year, to cover the lost interest. If we did, it would raise *Leiona*'s cost from £500 a year to £1,000, which is outside my range.

PERSONAL EXPENSES

These are much harder to consider than the boat expenses, as they depend even more on the individual.

As a general figure, either alone in the winter, when living is simpler and less ambitious, or with friends on board in the summer, I have found that about £4 a week for each person was enough for food and drink on board and ashore. This has been achieved by economic planning, and I am afraid that the cost is now increasing.

Prices of food and service have risen, as they have for labour on the boat, and so have house rents and hotel rates. These latter will not concern me, I hope, living aboard *Leiona* all year round. As a consideration, the rent of the little house I had at Puerto de Andraitx has doubled in the last three years from £8 a month to £16, and that is considered very cheap.

PROBLEMS

Sailing and living in a boat in the Mediterranean has a lot of pleasures, but it also has some problems. Luckily the good outweigh the bad, but the problems are always there, and they can be a shock to someone who is only used to sailing in British waters.

These problems again differ with the country, with the year, and with one's self, but some seem to be constant.

A general problem is that of non-availability of gear. Very often some vital material, on which the boat's running, economy, or safety depend either does not exist in the country of the moment, or has gone out of supply in the area. This has just happened in the case of a particular kind of engine oil which I and many others use. Maybe it will return into supply soon, or some substitute will arrive, but at the moment it is one of the problems with no apparent solution. Or no easy one.

Many yachts have gas stoves or refrigerators. It is not always easy to have the bottles refilled, and it is usually impossible to exchange them. Buying the local type, even if available, may involve changing the connections. And it is even possible to find that although the connections fit, the gas does not work in the equipment of another country.

Luckily, two problems which might seem awkward to the English reader never seem to produce any difficulties. They are the problems of entry into foreign countries, and of language barriers once there. Wherever I have gone I have always met with courtesy and helpfulness, and where there has been a language difficulty there has always been someone to help solve it.

I am sure that the pleasures of Mediterranean sailing are well worth its prices and problems.

Puerto de Andraitx. 1967

INDEX

76